# THE LOST WOMEN OF SCIENCE

A SERIES FROM THE LOST WOMEN OF SCIENCE INITIATIVE

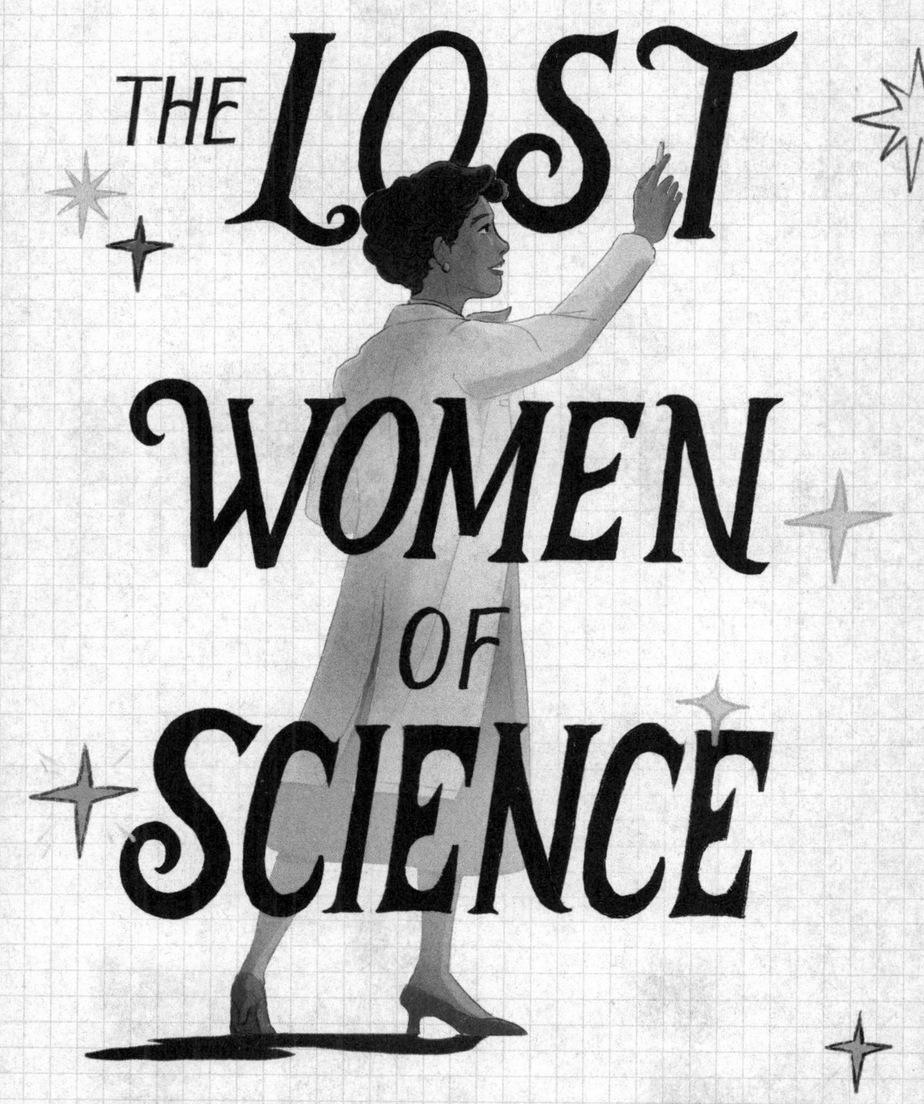

# THE LOST WOMEN OF SCIENCE

MELINA GEROSA BELLOWS & KATIE HAFNER

ILLUSTRATED BY KARYN LEE

BRIGHT MATTER BOOKS
NEW YORK

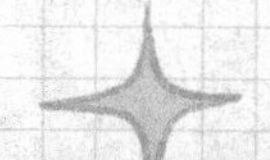

Text copyright © 2025 by Melina Gerosa Bellows and The Lost Women of Science Initiative
Cover art and interior illustrations copyright © 2025 by Karyn Lee

All rights reserved. Published in the United States by Bright Matter Books, an imprint of Random House Children's Books, a division of Penguin Random House LLC, 1745 Broadway, New York, NY 10019.

Bright Matter Books and the colophon are registered trademarks of Penguin Random House LLC.

penguinrandomhouse.com
rhcbooks.com

Library of Congress Cataloging-in-Publication Data
Names: Bellows, Melina Gerosa, author. | Hafner, Katie, author. | Lee, Karyn, illustrator.
Title: The lost women of science / Melina Gerosa Bellows & Katie Hafner; illustrated by Karyn Lee.
Description: First edition. | New York: Bright Matter Books, [2025] | Series: A series from the Lost Women of Science Initiative | Audience: Ages 10–12 | Audience: Grades 4–6 | Summary: "From the creators of award-nominated Lost Women of Science podcast comes an illuminating and moving portrayal of ten revolutionary women in STEM who dared to break barriers even when no one was watching. Through riveting sketches, rarely-before-seen photos, and insightful stories, unearth the lives and triumphs of these science pioneers whose influence cannot be forgotten"—Provided by publisher.
Identifiers: LCCN 2024049041 (print) | LCCN 2024049042 (ebook) | ISBN 978-0-593-89671-6 (trade) ISBN 978-0-593-89672-3 (lib. bdg.) | ISBN 978-0-593-89673-0 (ebook)
Subjects: LCSH: Women scientists—Biography—Juvenile literature.
Classification: LCC Q141 .B357 2025 (print) | LCC Q141 (ebook) | DDC 509.2/52—dc23/eng/20250111

The illustrations for this book were created digitally.
The text of this book is set in 11-point Maxime Pro.
Interior design by Carol Ly

Printed in the United States of America
10 9 8 7 6 5 4 3 2 1

The authorized representative in the EU for product safety and compliance is Penguin Random House Ireland, Morrison Chambers, 32 Nassau Street, Dublin D02 YH68, Ireland, https://eu-contact.penguin.ie.

Random House Children's Books supports the First Amendment and celebrates the right to read.

Penguin Random House values and supports copyright. Copyright fuels creativity, encourages diverse voices, promotes free speech, and creates a vibrant culture. Thank you for buying an authorized edition of this book and for complying with copyright laws by not reproducing, scanning, or distributing any part of it in any form without permission. You are supporting writers and allowing Penguin Random House to continue to publish books for every reader. Please note that no part of this book may be used or reproduced in any manner for the purpose of training artificial intelligence technologies or systems.

To my children, Chase and Mackenzie

—M.G.B.

To all the forgotten women whose stories we have told and those whose stories we hope yet to tell

—K.H.

# CONTENTS

# FOREWORD

Throughout our lives, we embark on journeys of all kinds. Some are physical. Others are adventures of the mind.

Our own journey at Lost Women of Science began in 2021, when we learned about Dr. Dorothy Andersen, who was born more than a hundred years ago and grew up to become a doctor. She wanted to be a surgeon, but during that time, women weren't welcomed into that medical specialty—nor into many others! So she became a pathologist, studying the causes of disease and death. As you'll learn in this book, Dr. Andersen discovered cystic fibrosis, a hereditary disease that was killing babies and young children. And as soon as she told the world about it in 1938, scientists began to develop treatments for prolonging the lives of CF patients. Today, thanks to Dr. Andersen's foundational work, people with cystic fibrosis can expect to live closer to normal lifespans.

There was just one problem: Dorothy Andersen got completely lost. After she died in 1963, her story died along with her. By the beginning of the twenty-first century, the world had completely forgotten who she was.

We started Lost Women of Science in order to tell Dorothy's story. But then a question occurred to us: How many other Dorothy Andersens were out there? How many dozens or even hundreds of remarkable female scientists have had their stories lost to history?

So our team at Lost Women of Science set out to research the lives and work of as many women as we could find—women who made huge contributions to science but for reasons of time and place and the lack of a Y chromosome didn't get the recognition they deserved during their lifetime.

The research we do can be challenging and often labor-intensive. In addition to the ten biographies, the book is packed with lots of bonus extras. Throughout its pages, you will find sidebars including science experiments on the greenhouse effect, a DIY book cipher, and a seaweed cake recipe. There are timelines on astronomy and code-breaking, mini biographies of famous people like Stevie Wonder and Harriet Tubman, and quick dives on topics ranging from the funky world of fungi to the special boxes built to bring moon rocks back to Earth.

Still, our journey has been like a luxury cruise compared to the free soloing done by our lost scientists. All we have to do is tell their stories. They had to work in environments that were far more hostile toward women than you'll find today—just so they could help humanity, whether by solving a medical riddle, studying the greenhouse effect, designing rockets for NASA, or carving out another scientific path entirely. Accomplishing any of these feats required equal measures of perseverance, curiosity, and guts.

In this first volume of Lost Women of Science, you'll get to know ten of these women, including Cecilia Payne-Goposchkin, who discovered what stars are made of and turned the world on its head; Yvonne Y. Clark, whose groundbreaking achievements as a Black female mechanical engineer earned her the moniker First Lady of Engineering; Elizebeth Smith Friedman, a master code-breaker who played a pivotal role in both world wars, though for many years, no one knew what she had done—not even her family; and Leona Zacharias, my own grandmother, who in the 1940s conducted research into an epidemic of blindness among premature newborns.

We hope that in learning more about each and every one of them, your curiosity will be piqued and you'll be inspired to carry out some remarkable work of your own. These once lost, now found women of science officially pass the baton of exploration and discovery to you, the next generation of mavericks.

—Katie Hafner, along with Melina Gerosa Bellows
and the Lost Women of Science team

# EUNICE NEWTON FOOTE

JULY 17, 1819–SEPTEMBER 30, 1888

“Science was one of those domains where women were struggling to be heard, and Foote is among the pioneers whose work paved the way toward acceptance.”

—Amara Huddleston, Climate.gov

| **CLAIM TO FAME** | The Mother of Climate Science |
|---|---|
| **WHY** | For her cutting-edge experiment that theorized about why Earth was getting warmer, advancing climate science |

**FIRSTS**

* First person to theorize the concept of the greenhouse effect
* First inventor to patent the thermostatically controlled cooking stove
* An attendee of the first women's rights conventions at Seneca Falls in 1848, which launched the fight for women's right to vote

Eunice Newton didn't grow up in a scientific family—but science ran in her blood. Her father, Isaac, was a distant relative of Isaac Newton. That's right, the famous physicist and mathematician who helped us understand gravity and motion, among other discoveries. Eunice was born in 1819, and her father was a cattle runner, or a rancher who raises cattle to be sold for meat. Eunice was born in Goshen, Connecticut, but when she was three years old, her father moved the family—her mother, Thirza, and her ten brothers and sisters—in a covered wagon to Bloomfield, a town in western New York State, with the hope of improving his job prospects. Unfortunately, job opportunities in Bloomfield weren't much better for Isaac, but it turned out to be a lucky move for Eunice.

New York between 1830 and 1860 was "the progressive dynamo of much of the United States," explains Sally Gregory Kohlstedt, a science historian and professor emerita at the University of Minnesota. The state was thriving with freethinkers in every domain, from religion to science to civil rights. Several cities contained stops along the Underground Railroad, the network of safe houses that aided Black people escaping slavery in the South, and communities were rife with discussions of women's rights. There were even dress reform activists who believed that more practical women's clothing should replace the ornate Victorian style, with its tightly laced corsets and bulky layers of petticoats. Everything from mixed-race marriages to new religions was open for discussion, and Eunice was exposed to this modern thinking.

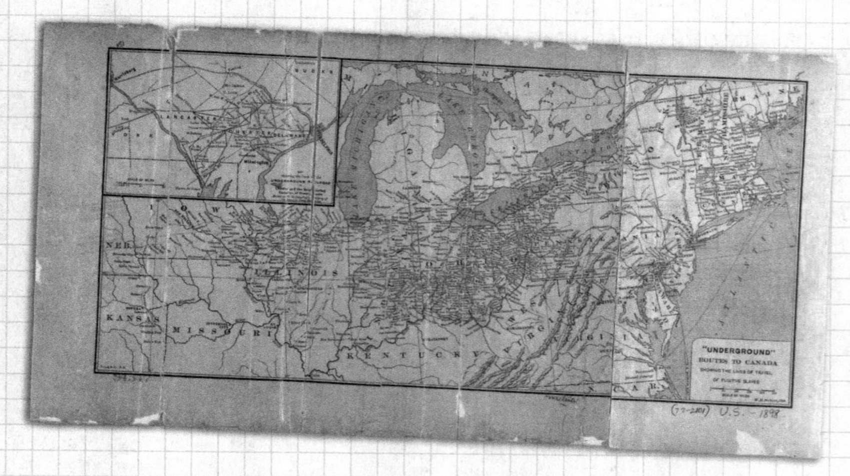


A map of the Underground Railroad marking the routes traveled to escape slavery.

Eunice's family were freethinkers too, and they invested in her education. They sent her to the feminist-founded Troy Female Seminary in Troy, New York. It was the first school in the country with a mission to provide young women with an education comparable to that of young college-educated men, a pretty revolutionary idea at the time. The seminary also happened to be right next to Rensselaer Polytechnic Institute, one of the most prestigious science schools in the country. Troy students could take classes there—although the women were still expected to be homemakers and not do much else after college. Eunice took classes at Rensselaer, and she expanded her knowledge of scientific theory and practice, which would have a big impact on the course of her life.

When Eunice's father died in 1835, he left his family with substantial debts. Soon the Newton farm was in danger of foreclosure, and the family had few options. To prevent the bank

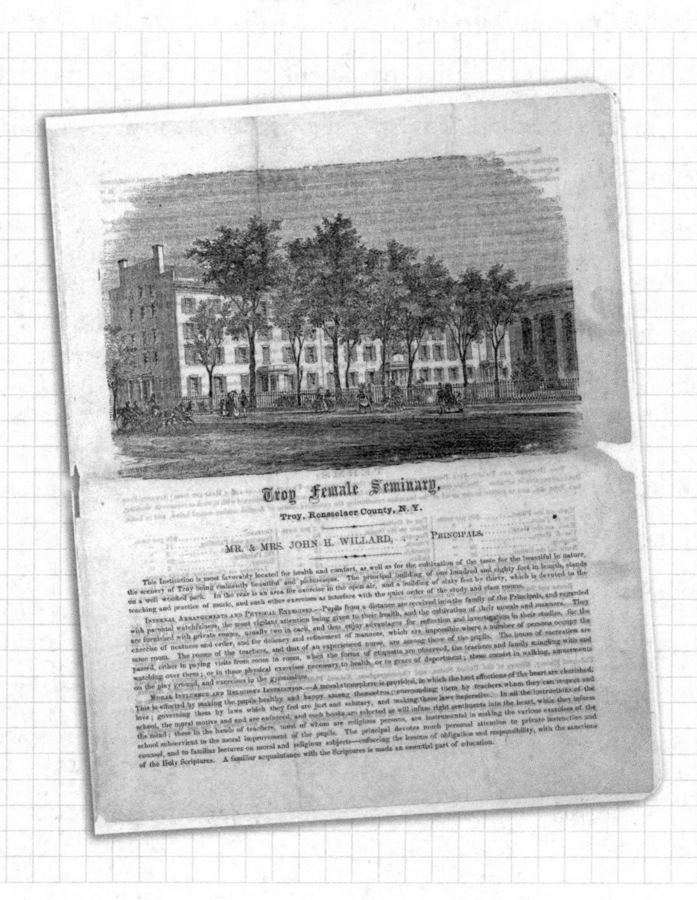

Troy Female Seminary,

Troy, Rensselaer County, N.Y.

MR. & MRS. JOHN H. WILLARD, PRINCIPALS.

This Institution is most favorably located for health and comfort, as well as for the cultivation of the taste for the beautiful in nature, the scenery of Troy being eminently beautiful and picturesque. The principal building of one hundred and eighty feet in length, stands on a well wooded park. In the rear is an area for exercise in the open air, and a building of sixty feet by thirty, which is devoted to the teaching and practice of music, and such other exercises as interfere with the quiet order of the study and class rooms.

INTERNAL ARRANGEMENTS AND PHYSICAL EXERCISES.—Pupils from a distance are received into the family of the Principals, and regarded with parental watchfulness, the most vigilant attention being given to their health, and the cultivation of their morals and manners. They are furnished with private rooms, usually two in each, and thus enjoy advantages for reflection and investigation in their studies, for the exercise of neatness and order, and for delicacy and refinement of manners, which are impossible where a number of persons occupy the same room. The rooms of the teachers, and that of an experienced nurse, are among those of the pupils. The hours of recreation are passed, either in paying visits from room to room, when the forms of etiquette are observed, the teachers and family mingling with and watching over them; or in those physical exercises necessary to health, or to grace of deportment; these consist in walking, amusements on the play ground, and exercises in the gymnasium.

MORAL INFLUENCE AND RELIGIOUS INSTRUCTION.—A moral atmosphere is provided, in which the best affections of the heart are cherished. This is effected by making the pupils healthy and happy among themselves; surrounding them by teachers whom they can respect and love; governing them by laws which they feel are just and salutary, and making these laws imperative. In all the instructions of the school, the moral motive and end are enforced, and such books are selected as will infuse right sentiments into the heart, while they inform the mind; these in the hands of teachers, most of whom are religious persons, are instrumental in making the various exercises of the school subservient to the moral improvement of the pupils. The principal devotes much personal attention to private instruction and counsel, and to familiar lectures on moral and religious subjects—enforcing the lessons of obligation and responsibility, with the sanctions of the Holy Scriptures. A familiar acquaintance with the Scriptures is made an essential part of education.

from taking the farm, Eunice's older sister Amanda jumped into action and hired the district attorney of the nearby town of Seneca Falls. The young lawyer Elisha Foote took on their case, won it, and also won the attention of Eunice.

Eunice and Elisha began a courtship, and in 1841 they married. She was twenty-two years old. He was thirty-two. The couple moved to Seneca Falls, and Eunice once again found herself in a town brimming with ideas. Seneca Falls was the center of the American women's rights movement. Eunice was in the right place at the right time, and most importantly, her open mind allowed her to jump into the action. In fact, one of the Footes' neighbors was none other than Elizabeth Cady Stanton, a leader of the suffrage movement fighting for women's right to vote. In 1848, Elizabeth coorganized the country's first women's rights convention in Seneca Falls, and Eunice and Elisha were right there with her, making history.

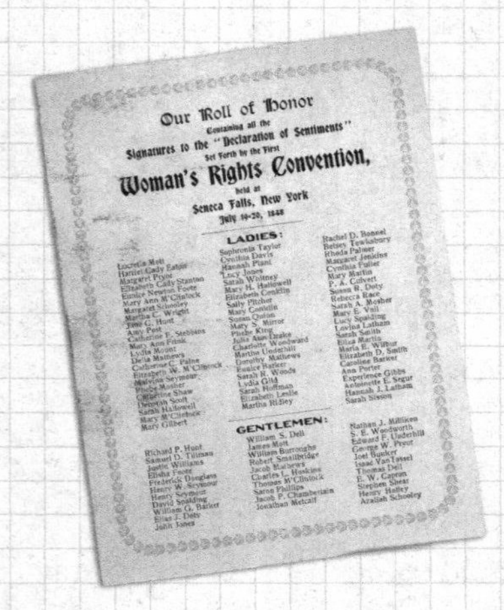
Our Roll of Honor
Containing all the
Signatures to the "Declaration of Sentiments"
Set Forth by the First
Woman's Rights Convention,
held at
Seneca Falls, New York
July 19-20, 1848

LADIES:

Lucretia Mott
Harriet Cady Eaton
Margaret Pryor
Elizabeth Cady Stanton
Eunice Newton Foote
Mary Ann M'Clintock
Margaret Schooley
Martha C. Wright
Jane C. Hunt
Amy Post
Catherine F. Stebbins
Mary Ann Frink
Lydia Mount
Delia Mathews
Catherine C. Paine
Elizabeth W. M'Clintock
Malvina Seymour
Phebe Mosher
Catherine Shaw
Deborah Scott
Sarah Hallowell
Mary M'Clintock
Mary Gilbert
Sophronia Taylor
Cynthia Davis
Hannah Plant
Lucy Jones
Sarah Whitney
Mary H. Hallowell
Elizabeth Conklin
Sally Pitcher
Mary Conklin
Susan Quinn
Mary S. Mirror
Phebe King
Julia Ann Drake
Charlotte Woodward
Martha Underhill
Dorothy Mathews
Eunice Barker
Sarah R. Woods
Lydia Gild
Sarah Hoffman
Elizabeth Leslie
Martha Ridley
Rachel D. Bonnel
Betsey Tewksbury
Rhoda Palmer
Margaret Jenkins
Cynthia Fuller
Mary Martin
P. A. Culvert
Susan R. Doty
Rebecca Race
Sarah A. Mosher
Mary E. Vail
Lucy Spalding
Lovina Latham
Sarah Smith
Eliza Martin
Maria E. Wilbur
Elizabeth D. Smith
Caroline Barker
Ann Porter
Experience Gibbs
Antoinette E. Segur
Hannah J. Latham
Sarah Sisson

GENTLEMEN:

Richard P. Hunt
Samuel D. Tillman
Justin Williams
Elisha Foote
Frederick Douglass
Henry W. Seymour
Henry Seymour
David Spalding
William G. Barker
Elias J. Doty
John Jones
William S. Dell
James Mott
William Burroughs
Robert Smallbridge
Jacob Mathews
Charles L. Hoskins
Thomas M'Clintock
Saron Phillips
Jacob P. Chamberlain
Jonathan Metcalf
Nathan J. Milliken
S. E. Woodworth
Edward F. Underhill
George W. Pryor
Joel Bunker
Isaac Van Tassel
Thomas Dell
E. W. Capron
Stephen Shear
Henry Hatley
Azaliah Schooley

This souvenir card recognizes the women and men who signed the "Declaration of Sentiments," a document that declared that women should hold equal stature with men, at the first women's rights convention at Seneca Falls. Eunice Foote's name can be seen in the first column.

"[Eunice] was immersed in a world that accepted her, that gave her self-confidence, I think, and that took her seriously," says Sally Gregory Kohlstedt. And perhaps Eunice's biggest champion was Elisha. He encouraged her interests and endeavors, which wasn't always a given in marriages during this period. They had a family together, raising two daughters. And they nurtured each other's endless curiosity, especially about all things science. Both Eunice and Elisha were inventors and collaborators. They were so inquisitive, they built themselves a home laboratory.

One of the questions puzzling Eunice was Earth's fluctuating temperature. In the mid-1800s, scientists already knew that Earth's temperature seemed to have seesawed over its history. But they weren't sure why. In 1856, Eunice set up a simple home experiment to look at how different gases trapped heat. First, she collected glass cylinders and placed thermometers inside each one. Then she filled the cylinders with different types of gases: regular air, carbon dioxide, dry air (air without a lot of moisture), and humid air (air with plenty of moisture). Lastly, she positioned some of the cylinders in the sun to be warmed and some in the shade to stay cooler.

Eunice noticed a few curious things: When exposed to sunlight, the humid air got hotter than the dry air, and oxygen heated up slightly more than hydrogen. But the biggest difference Eunice noted was between regular air and carbon dioxide.

In the sun, a tube of regular air heated to 100°F, while carbon dioxide shot up to 120°F.

That's when Eunice's basic science experiment blossomed into a big idea: What if there were times when the Earth's atmosphere had more carbon dioxide in it? Could changing carbon dioxide levels cause the fluctuations in Earth's temperature? In a short article published in 1856, she posed this idea: "An atmosphere of that gas would give to our earth a high temperature."

These days, scientists know that the atmosphere is a mix of gases, mostly nitrogen. Carbon dioxide makes up a tiny proportion of it. But Eunice concluded that even a small change in the amount of carbon dioxide in the air could shift the temperature of the entire planet. She wrote a paper proposing that variations in carbon dioxide could explain why Earth had been warmer or colder at different points in history. Eunice's bottom line: More carbon dioxide meant a warmer climate. It was the first research to propose the concept of the greenhouse effect, although Eunice did not call it that at the time.

## Greenhouse Gases

Life on Earth would not be possible without greenhouse gases. Carbon dioxide, methane, nitrous oxide, and water vapor are all critical parts

of the planet's atmosphere because they trap heat from the sun. This trapped light energy creates a warm, life-sustaining environment. Without these protective gases, humans would freeze to death.

Just as the name suggests, the phenomenon works similarly to a greenhouse. The sun radiates light to our planet (our greenhouse). Some of that light is reflected back into space by Earth's surface, much as sunlight bounces off a greenhouse. But some of the light is trapped by greenhouse gases, which creates a large blanket of air wrapping around our planet to keep it warm.

That may sound cozy, but human activity these days is producing an excessive amount of greenhouse gases. Burning fossil fuels (such as coal, oil, and natural gas) for electricity, heat, and transportation is creating too much hot air, which gets trapped in our atmosphere. Trees and plants, which absorb $CO_2$ and convert it into oxygen through photosynthesis, can act as critical carbon sponges, absorbing the gas from Earth's atmosphere. However, deforestation means there simply isn't enough vegetation to keep up with the amount of carbon released. Oceans also absorb about a quarter of the $CO_2$ we release into the atmosphere each year, but this absorption is leading to other issues like ocean acidification. The water is becoming more acidic, harming marine life like corals and shellfish.

As a result of this carbon excess, Earth's temperature is rising, a phenomenon known as global warming, and it's a key indicator of climate change. Ice caps and glaciers are melting, wildlife populations and habitats are negatively impacted, and hurricanes and other severe weather events are becoming more frequent.

Reducing greenhouse gas emissions will take a worldwide collaboration of governments, industries, communities, and individuals. You can do your part. Bike to school or use public transportation. Reduce, reuse, and recycle whenever possible. Planting trees and educating others about what you know can make a difference.

"As we now know, climate change is caused by heat-trapping gases building up in the atmosphere, essentially wrapping an extra blanket around the planet," says Katharine Hayhoe, a climate researcher and chief scientist at the Nature Conservancy. "Here she was in the 1850s, clearly explaining that to the scientists of the day."

# Demonstrating the Greenhouse Effect

Feeling the curiosity itch like Eunice? You can create your own home experiment to see the greenhouse effect in action. Be sure to find a trusted adult to help when setting up the heat lamp or light bulbs. You can even call them your lab assistant!

**You will need:**

* Two identical clear glass jars
* Two thermometers
* Plastic wrap
* Rubber bands
* Two small heat lamps or two lamps with incandescent bulbs
* A timer

**Procedure:**

1. Place a thermometer inside each of the two jars.
2. Cover the opening of one jar with plastic wrap and secure it tightly with the rubber bands. Leave the other jar open.
3. Position the light bulbs or heat lamps at equal distance above each jar, being mindful that the lamps don't directly touch the plastic wrap. The lamps will replicate sunlight.
4. Record the initial temperature in both jars and note the time.
5. Every five to ten minutes for the next hour, record the temperature inside each jar.

**Observation:**

When you review your results, you'll likely see the temperature in both jars rise. However, the jar with the plastic wrap will probably be warmer than the one without. That's because the plastic wrap acts similarly to Earth's atmosphere. It traps the light energy from the heat lamp or light bulb and keeps the warmed air inside the jar, raising the overall temperature. Earth's greenhouse gases do the same thing.

Eunice submitted her findings to the American Association for the Advancement of Science (AAAS), one of the country's first national science associations. It would have been unconventional for a woman in the 1850s to present the material—even if it was her own—so she asked a man to do it for her. Physicist Joseph Henry was the secretary of the Smithsonian Institution, the leading scientific organization in the country. At its annual convention, Joseph opened the presentation by acknowledging that the research about to be presented was the work of a female scientist. However, few seemed to take notice of the obligatory mention, and Eunice watched from her seat as Joseph presented her findings.

Although Eunice's paper didn't make it into the official conference proceedings, she formally published it a few months later in the *American Journal of Science and Arts,* where it made a splash in the scientific community. She got a write-up in the *Annual of Scientific Discovery* and a glowing review in the magazine *Scientific American*. But the excitement was short-lived. Eunice returned to life with Elisha and their children, and her academic papers were filed away.

Around the time that Eunice began questioning the role of carbon dioxide in Earth's temperature, an Irish scientist named John Tyndall started looking into similar questions. In the late 1850s, John also conducted an experiment with gases, heat, and thermometers at the Royal Institution in London. But unlike Eunice with her home lab, John had significantly more resources to work with, including the most modern equipment available, as well as lab assistants. These advantages enabled him to take the research one step further. Eunice had demonstrated the greenhouse gas effect, but she could not determine *why* some gases heated up so much more than others.

382 *On the Heat in the Sun's Rays.*

ART. XXXI.—*Circumstances affecting the Heat of the Sun's Rays*; by EUNICE FOOTE.

(Read before the American Association, August 23d, 1856.)

MY investigations have had for their object to determine the different circumstances that affect the thermal action of the rays of light that proceed from the sun.

Several results have been obtained.

First. The action increases with the density of the air, and is diminished as it becomes more rarified.

The experiments were made with an air-pump and two cylindrical receivers of the same size, about four inches in diameter and thirty in length. In each were placed two thermometers, and the air was exhausted from one and condensed in the other. After both had acquired the same temperature they were placed in the sun, side by side, and while the action of the sun's rays rose to 110° in the condensed tube, it attained only 88° in the other. I had no means at hand of measuring the degree of condensation or rarefaction.

The observations taken once in two or three minutes, were as follows:

| Exhausted Tube | | Condensed Tube. | |
|---|---|---|---|
| In shade. | In sun. | In shade. | In sun. |
| 75 | 80 | 75 | 80 |
| 76 | 82 | 78 | 95 |
| 80 | 82 | 80 | 100 |
| 83 | 86 | 82 | 105 |
| 84 | 88 | 85 | 110 |

This circumstance must affect the power of the sun's rays in different places, and contribute to produce their feeble action on the summits of lofty mountains.

Secondly. The action of the sun's rays was found to be greater in moist than in dry air.

In one of the receivers the air was saturated with moisture—in the other it was dried by the use of chlorid of calcium.

Both were placed in the sun as before and the result was as follows:

| Dry Air. | | Damp Air. | |
|---|---|---|---|
| In shade. | In sun. | In shade. | In sun. |
| 75 | 75 | 75 | 75 |
| 78 | 88 | 78 | 90 |
| 82 | 102 | 82 | 106 |
| 82 | 104 | 82 | 110 |
| 82 | 105 | 82 | 114 |
| 82 | 108 | 92 | 120 |

A page from Eunice's paper published in the *American Journal of Science* in 1856.

Instead of putting his gases in the sun, John used a copper tube filled with boiling water. Like any hot object, the tube gave off radiant heat—which scientists now call long-wave infrared radiation. Greenhouse gases are extra effective at absorbing radiated heat. John was able to measure how much radiation the gases were absorbing by using a device called a spectrometer that he built himself. He also showed that sunlight could easily pass through gases. Like Eunice, he later wrote that changing the concentrations of these gases could explain the fluctuating temperatures of the planet. This time, though, the scientific community heralded the theory as groundbreaking. In the centuries since John's research, he has even been called the Father of Climate Science.

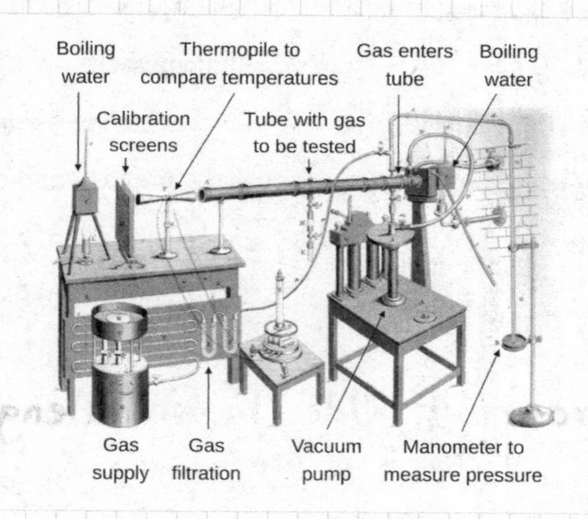


A depiction of John Tyndall's experiment measuring how different gases absorb radiant heat.

Did John Tyndall know about Eunice Foote's work before he conducted his research? Historians have mixed views. John Perlin, who teaches physics at UC Santa Barbara and is writing a book about Eunice, believes that what is apparent is that they were interested in the same subject. There is also evidence that John Tyndall was editing a magazine that contained a reprinted article by Elisha Foote. The article had originally appeared next to Eunice's paper. So if John had seen Elisha Foote's original article and decided to republish it . . . would he have seen Eunice's, too?

It's an intriguing question to ponder. Especially considering that John Tyndall was suspected of other plagiarism. He was also accused of blatantly stealing credit for research on sound waves from Joseph Henry, the scientist who presented Eunice's work at the AAAS conference. To others, the case is less clear-cut. "On the other hand, in the history of science, there's a lot of what we call simultaneous discoveries," says Sally Gregory Kohlstedt. "Sometimes at two different places in two different ways, two scholars do the same thing."

## Spectroscopy: Ride the Wave(length)!

When Eunice had the bright idea to investigate how different gases absorb light, she was viewing the visible light spectrum. John

Tyndall's experiments in the 1850s and 1860s, however, were able to look at the infrared spectrum—or light that humans cannot see with the naked eye.

John was studying a branch of science called spectroscopy, which examines how matter interacts with light. When light waves hit an object, the light can be reflected, absorbed, or transmitted. By studying light's effects, scientists learn a lot about the objects themselves.

Consider a rainbow. When light passes through a prism—such as raindrops—it breaks white light into the spectrum of colors. Every object emits light, but not all light can be seen without the help of an optical instrument such as a telescope, a microscope, or, in this case, a spectrometer. A spectrometer breaks up the different light waves an object emits into atoms, or tiny units of matter. These atoms are then translated into visible light and given their own color so that humans can see and identify them.

Because different substances absorb light differently, scientists can observe the changes and amounts of colors, which are often called spectra, to determine chemical composition, molecular structure, and the physical properties of the objects they are studying.

Spectroscopic techniques are used in multiple scientific fields, including chemistry, physics, and astronomy. Scientists use spectroscopy for everything from examining microscopic amounts of blood in a laboratory to viewing gigantic stars in distant galaxies. Spectroscopy illuminates compelling research in climate science!

Whether plagiarism or coincidence, John Tyndall received all the credit and became widely known. Eunice, meanwhile, faded from public view. She focused on family life and continued inventing into her forties. In 1842, she and Elisha had patented a thermostatically controlled cooking stove, which made it possible to set ovens to more precise temperatures. And in 1857, they won a hefty court settlement when someone else tried to steal their invention. In 1860, she filed a patent in her own name on a rubber shoe insert that was intended to "prevent the squeaking of boots and shoes." She also developed a new type of paper-making machine that lowered the cost of manufacturing. Her curious mind never stopped searching for answers and solutions.

Eunice Newton Foote died at age sixty-nine in 1888, a few years after Elisha. For more than a century, she was almost entirely forgotten. Then, in 2011 Ray Sorenson, a collector of antique science publications, found a volume of the *Annual of Scientific Discovery*. The magazine recapped each year's biggest scientific highlights, and he read about Professor Joseph Henry presenting Eunice's paper on the greenhouse effect at the AAAS.

"If she's the first one to do this, she needs to be given credit for it," says Ray. He wrote a paper clarifying and paying tribute to Eunice's early contribution, hoping to entice researchers to dig up more about this long-lost trailblazer. Not only did Ray's paper generate interest, but the scientists who examined Eunice's findings also confirmed that Ray was onto something. Eunice was not able to isolate the impact of infrared radiation, as John Tyndall's experiments did. However, her assertion that water and carbon dioxide absorb heat from sunlight and that this property could result in climate change long-term predated John Tyndall's discovery by three years. In 2022, 134 years after her death, the American Geophysical Union created the Eunice Newton Foote Medal for Earth-Life Science in her honor.

Glory, however, was never the point. Eunice was driven by a relentless passion for discovery. "Ultimately, my assumption is that she followed her own instincts," says Sally Gregory Kohlstedt. "She wanted to make a contribution to knowledge." Eunice grew up surrounded by ideas, and by the end of her life, she had contributed many of her own. Her distant relative Isaac Newton would have been proud.

# FLORA PATTERSON

SEPTEMBER 15, 1847–FEBRUARY 5, 1928

"With a name like Flora, she seemed destined for a master's in botany, the study of plants."

—Hilda Gitchell,
*Lost Women of Science* producer

| **CLAIM TO FAME** | Mycologist in Charge |
|---|---|
| **WHY** | She prevented the spread of numerous diseases that could threaten plant life in the United States. |

**FIRSTS**

* First female mycologist (fungus expert) and plant pathologist at the US Department of Agriculture
* First person to identify a deadly fungus that caused the destruction of the American chestnut tree
* First person to implement a screening system for plant material entering the US from foreign countries

Flora Wambaugh Patterson's appreciation for nature was seeded at a young age. However, as with many plants (and people), it took time for her skills to bloom. Flora was born in 1847 in Columbus, Ohio. At an early age, she showed an interest in learning about fungi—which include mushrooms—as a hobby.

In her later schooling, Flora earned her bachelor's degree at the Cincinnati Wesleyan College for Young Women in Ohio, graduating in 1865. The Civil War had just ended, and this was a time when only about one in every six college graduates was a woman, so having a university degree in itself was an accomplishment.

But Flora wanted more. She went on to earn a master's degree from Wesleyan University in 1883. Despite her academic accomplishments, though, Flora's job prospects were slim. Four years after finishing college, she married Edwin Patterson of Ripley, Ohio, and they settled in nearby Cincinnati, where he was a steamboat pilot. She moved away from academia and focused on family life, giving birth first to one son, followed by another soon after. Then tragedy struck. Her husband was badly injured in a steamboat explosion. His injuries meant that he required Flora's attentive care, which she provided for ten years, until he died.

Now without a husband, Flora needed to figure out a plan to support herself and her two kids. So she moved her family from Ohio to Iowa, where her brother was a professor at the University of Iowa (called the State University of Iowa at the time). In Iowa City, Flora returned to her roots (literally), enrolling in classes to get her master's degree in botany, or the study of plants. Just as Flora finished her coursework, her brother took a new post at Harvard University in Massachusetts. Flora followed him, intending to study at nearby Yale. However, only after moving did a dismayed Flora learn that Yale was not accepting women into their botany program. Instead, Flora would take classes at Radcliffe College, the all-women's sister school of Harvard. There she got a job at the Gray Herbarium. An herbarium is like a library for dried plants, and Flora's job included inspecting, labeling, and packaging fungi.

Officially named in 1893, the Gray Herbarium houses a diverse array of specimens on the Harvard University campus.

Flora's official job title was assistant, but that label fell short in describing her growing expertise. After three years honing her skills, she was able to both preserve a specimen and identify it by sight. Flora was an expert . . . but she was also a woman, and could not advance any further in her current job. So in 1895, twenty-five years before women got the right to vote, she did something clever to get the job she wanted: She took a civil service examination, which was a test to get a job in the United States government. And Flora aced it.

The exam catapulted the forty-eight-year-old widowed mother of two to a job at the US Department of Agriculture, the USDA. Flora worked hard in this new role. Her skills were appreciated, even by those who resisted working alongside women. When Flora began her career with the USDA, about 15 percent of the workers were women. However, most of those women were secretaries or typists or worked in other clerical roles. Flora was unique in that she was one of the few women able to immerse herself in the fungal collection. "It's like a library of fungi," says Amy Rossman, a mycologist at the USDA who has researched Flora's life.

And it's no wonder Flora had a fascination with fungi; they're incredible! Far from simply a funk on your feet, a fungus is a living thing that is neither a plant nor an animal. Fungi are both helpful and harmful. They're critical for life on Earth, trapping carbon in the soil. Penicillin, a vaccine derived from a fungus, has saved countless lives. And mushrooms, also fungi, are a healthy pizza topping. But if a new fungus enters an ecosystem, it can spell disaster for the existing plant life. The natural vegetation has no existing defense against the new intruder, making it vulnerable to harm.

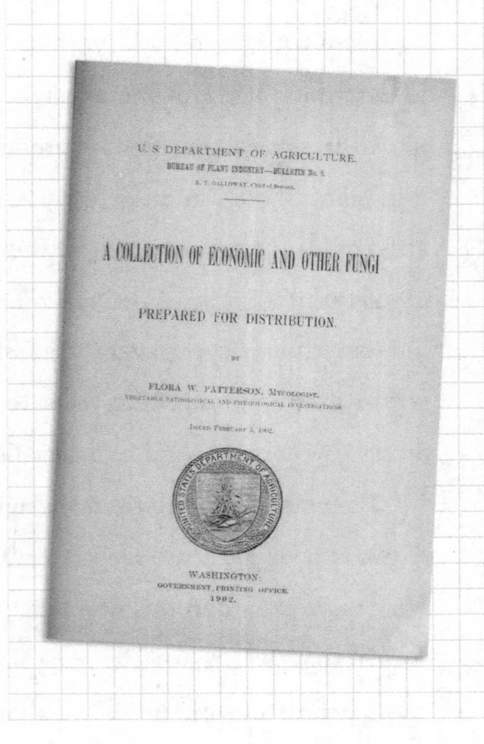
U. S. DEPARTMENT OF AGRICULTURE.
BUREAU OF PLANT INDUSTRY—BULLETIN No. 8.
B. T. GALLOWAY, Chief of Bureau.

A COLLECTION OF ECONOMIC AND OTHER FUNGI

PREPARED FOR DISTRIBUTION.

BY

FLORA W. PATTERSON, MYCOLOGIST,
VEGETABLE PATHOLOGICAL AND PHYSIOLOGICAL INVESTIGATIONS.

ISSUED FEBRUARY 3, 1902.



WASHINGTON:
GOVERNMENT PRINTING OFFICE.
1902.

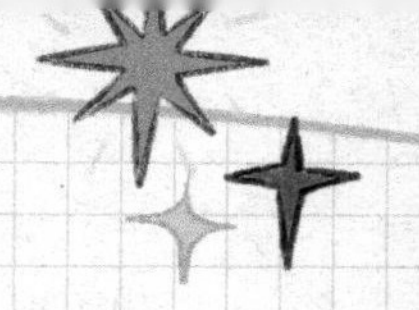

# The Funky World of Fungi

With more than 144,000 known species, fungi are as weird as they are wondrous. The fungus family is made up of mushrooms, molds, yeasts, mildews, and more. Fungi act more like animals than plants because of the way they reproduce, as well as their inability to make their own food (unlike plants).

Mushrooms sprouting in the forest make up just a fraction of the fungus world. Fungi come in various shapes, colors, and sizes. In fact, the largest organism on Earth is a honey fungus in eastern Oregon's Malheur National Forest. Not only is it almost four square miles in size, but it's also potentially 8,650 years old. Talk about an old mold! Fungi have been known to survive in diverse ecosystems such as frigid polar regions, steamy rainforests, and even deep beneath the ocean floor, where few other organisms can survive—let alone thrive—in the pitch black.

Fungi also play important roles in our daily lives. The mold in yeast is what makes bread rise, mold gives blue cheese its color, and mold is what makes soy sauce so tasty. Perhaps the most famous fungus is *Penicillium*, which helped create the lifesaving medicine penicillin, the first mass-produced antibiotic. However, *Penicillium* is not the only medicinal fungus. *Taxomyces* has been used in medicines to fight cancer.

Some fungi are capable of amazing feats. The jack-o'-lantern

mushroom, found in North America, Africa, and Europe, is bioluminescent and emits a greenish glow. A fungus called *Cordyceps* infects insects, taking over their bodies and making them like zombies. The stinkhorn mushroom exudes a smelly odor that cleverly attracts flies, which help the mushrooms distribute their spores.

The jack-o'-lantern mushroom is known for its ability to glow, called bioluminescence.

Mycorrhizal fungi even have the ability to form mutually beneficial relationships with other organisms. Beneath the ground, an intricate network of fungal threads called mycelium creates a symbiotic relationship with plants. As the fungi provide plants with vital nutrients, they gain sugars they need for energy.

Fungi are also talented recyclers, adept at decomposing dead organic matter and turning it into rich soil. This ability has prompted scientists to explore the potential for using certain fungi to help with environmental problems. *Pestalotiopsis microspora*, for example, is a fungus discovered in the Amazon rainforest that is capable of breaking down polyurethane, a common type of plastic, through a process called biodegradation. White-rot fungi can absorb heavy metals like mercury, lead, and arsenic, thanks to their unique enzyme system, which allows them to break down toxic substances.

Researchers have only begun to unlock the potential of fungi. One thing is certain, though; fungi are sure to grow on you.

In 1904, the American chestnut tree encountered such a disease, which came to be called chestnut blight. First, tiny red spots appeared on the tree's trunk and leaves. Then the blight would produce spores that got caught in the wind and traveled to other trees. Soon entire forests were wiped out.

At first, the fungal blight took hold quietly. A forester at New York City's Bronx Zoo was the first to notice that a mighty American chestnut tree was killed by a small fungal infection that was barely perceptible to the untrained eye. But the little red spots clumped on the trees were deadly—and spreading rapidly. The blight traveled quickly as far west as Ohio. As it did, Flora dug deep into the USDA's fungal collection. By now, Flora was considered an expert in the field and therefore became the go-to person to locate the source of this new tree infection. And they called the right woman for the job. Flora is credited with being the first to identify *Cryphonectria parasitica* as the lethal fungus responsible for taking down so many trees. It is speculated that the fungus likely originated when Japanese chestnuts reached American shores.

Even with Flora's useful identification, it was too late to save the trees. Within fifty years, the previously towering American chestnuts were gone. For Flora, the worry was far from over. She knew that if such a contamination could happen once, it would happen again unless safety checks were put in place. Her solution? Plant imports arriving at the US border should be inspected. And Flora urged authorities to act fast, because blight could be right around the corner, threatening not just trees, but food sources as well.

Five years later, in 1909, Flora's premonition nearly came true. Japan had sent 2,000 cherry tree saplings to Washington, DC, as a goodwill gift. The trees, which were to be planted all along the tidal basin near the White House, traveled for weeks by boat from Japan to Seattle, and then crossed the country on trucks to DC.

By the time the trees got there, Flora and her team of inspectors were ready to greet them. It was a good thing, too, as the team discovered that the trees were covered with fungi and insects. Flora had to deal with the embarrassing problem of rejecting the gift from Japan. Per her team's recommendation, the Department of Agriculture ended up burning trees on DC's National Mall in a bonfire right in front of the Washington Monument.

After the destruction of the original trees, a second shipment was sent. Fortunately, Flora gave these trees a clean bill of health, and they were planted in Washington. To this day, millions of tourists come every spring to see their pink splendor.

The public outcry over the burning of the cherry trees and the chestnut blight five years earlier brought public attention to the serious threat and the astronomical cost of invasive pests, launching a national discussion. Flora doubled down on her push for a federal policy to stop plants at the border so they could be checked for disease.

"She was able to write very convincingly that these things were coming into the country and had to be stopped," says Dr. Sandra Anagnostakis, a scientist studying chestnut blight disease.

"And it was because of her work . . . that the Plant Quarantine Act was finally passed in 1912." This act mandated the inspection of imports and established border checks to prevent infectious plants from spreading another plant pandemic.

However, merely establishing inspections was only part of the challenge. For the inspections to be effective, employees needed to be able to identify invasive fungi precisely. And fungus species outnumber plant species by at least six to one. Once again, Flora put her head down and got to work. She was able to expand the USDA's fungal collection, enabling inspectors to know just where to look if they needed to identify anything suspicious.

## Paw Patrol

In addition to Flora, we also have canines to thank for our national biosecurity. The USDA uses specially trained dogs for various purposes, including detecting plants and insects at ports of entry throughout the country. The paw patrol's job is to prevent the introduction of harmful pests, diseases, and invasive species. These threats could jeopardize our country's food crops, forests, farms, and environment—and the livelihoods of America's farmers and ranchers.

The National Detector Dog Training Center started with the Beagle Brigade in 1984, when two dogs, Canine Bucky and Canine

Lady, were used at the Los Angeles International Airport to detect plants and animal products in carry-on bags and luggage arriving from foreign destinations. Since then, the center has grown into an extensive operation that trains more than 5,000 dogs on their Newnan, Georgia, campus for numerous federal agencies, including the USDA.

After graduating from rigorous training to develop their exceptional olfactory (smelling) senses, these highly effective teams inspect passenger baggage, cargo, and parcels to detect fruits, vegetables, and meats in international passenger baggage, mailed packages, and vehicles entering the United States.

The dog breeds selected have a number of attributes, including high energy and intelligence, the ability to focus, and the desire to work hard. Beagles in particular can detect the scents emitted by plant pests and diseases that humans could never detect. The center trains Jack Russell terriers to stop brown tree snakes—which have caused the extinction of several bird species in Guam—from reaching Hawaii and the Mariana Islands. The center also trains Labrador retrievers to detect nutria, invasive rodents that destroy wetlands on Maryland's Eastern Shore.

In addition to identifying exotic fruits and vegetables that travelers attempt to smuggle in, sniffer dogs can also detect bedbugs, narcotics, explosives, firearms, poached wildlife, ivory and other items made from endangered animals, and unusual meats like dried caterpillars and kangaroo.

The dogs' natural scent recognition enables them to accurately identify suspicious items up to 85 percent of the time. They play a crucial role in protecting US agriculture, conservation, food security, and the economic stability of farmers. Talk about a job done doggone well!

During Flora's twenty-seven years at the USDA, the fungal collection grew to almost 115,000 specimens, over five times the size it was when she first started working there in 1896. Eight hundred of those specimens were added by Flora herself. This is extraordinary, considering that preparing an individual specimen is a lengthy and painstaking process.

Still, even though Flora and her four employees (three of whom were women) had won accolades for helping implement the 1912 Quarantine Act, Flora was notified that she and her team would be transferred to a department where they would no longer be inspecting imported plants.

The change was presented as a cost-cutting measure. But Flora wasn't going down without a fight. She sent a strongly worded letter to her supervisors, reminding them that her group of inspectors had already prevented the spread of several potential diseases, writing, "It is significant that each fungus disease which has been called to public attention through the Department or by other workers had first been noted by the inspectors either directly or by means of correspondence." Flora's letter went on to list several times when her mycological team had found dangerous plant infections before they were able to spread, including English potato scab, silver scurf, chestnut blight disease, and citrus canker.

Flora won that battle, and she stayed on the job for seven more years, retiring at age seventy-five. There is no evidence that Flora remarried. She had a full career, and she was living with one of her sons in New York City when she died in 1928, at the age of eighty, five years after she retired.

Since then, a long line of women have followed in Flora's footsteps as lead mycologists at the USDA. But perhaps Flora's greatest legacy is the invaluable US National Fungus Collections itself. The collection that Flora painstakingly helped build is still the largest in the world. And it is her work that allows it to continue to grow.

# DR. SARAH LOGUEN FRASER

JANUARY 29, 1850–APRIL 9, 1933

“I will never, never see a human being in need of aid and not be able to help.”

—Dr. Sarah Loguen Fraser

| CLAIM TO FAME | Miss Doc |
|---|---|
| WHY | One of the first Black female doctors in the United States and the first female doctor in the Dominican Republic |

**FIRSTS**

* First Black MD to graduate from a coed medical school
* First female doctor in the Dominican Republic

When Sarah Loguen Fraser was a little girl, one man, two women, and six children arrived at her home in Syracuse, New York. They were a group of enslaved people—and they all had gunshot wounds to their legs. Little Sarah jumped into action alongside her mother. Their home was a stop on the Underground Railroad—a network of people who created safe houses and trails to aid Blacks escaping the South—and they were ready to help. Sarah bathed the wounded legs of one of the girls. Later, she would recall this powerful experience and say it made her feel like "she was 'the most important person in the whole house,'" according to an unpublished biography by Sarah's daughter, Gregoria Fraser Goins.

Sarah Loguen was born in 1850 in Syracuse, a city near the center of New York State. The fifth of eight children, Sarah was the daughter of escaped slave Reverend Jermain Wesley Loguen and his wife, Caroline. Her parents were abolitionists, people who were opposed to slavery and joined a movement to abolish it.

Unfortunately, not everyone agreed with them. The same year Sarah was born, the United States Congress passed the Fugitive Slave Act. This act allowed federal marshals to track down and capture people who had escaped from the slaveholding Southern states and return them to bondage. Anyone who tried to help these individuals, like Sarah's parents, could be charged with a federal crime and sent to jail. Her father, as a formerly enslaved person, also risked reenslavement.

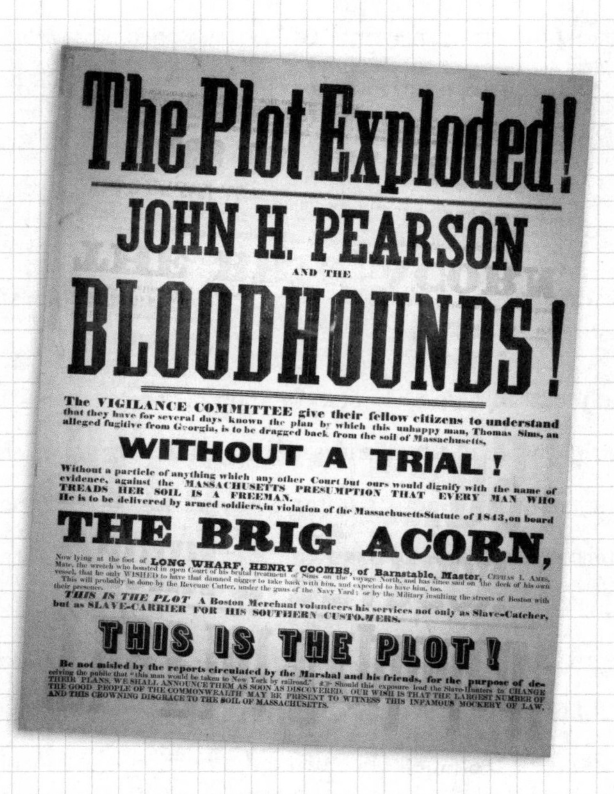


This poster advocates for the release of John H. Pearson, an enslaved seventeen-year-old from Georgia who escaped to Boston. Pearson was captured under the Fugitive Slave Act of 1850 and sold back into slavery. He was able to escape again and finally find freedom in Boston.

Syracuse was an active place during the abolitionist movement, and Sarah's father was known as the King of the Underground Railroad. He had brave and powerful friends, including Underground Railroad conductor and social activist Harriet Tubman and abolitionist and orator Frederick Douglass. Sarah's parents worked side by side at their station on the Underground

Railroad, protecting an estimated 1,500 escaped enslaved people, knowing they risked terrible punishments if caught. And for the first decade of her life, Sarah witnessed all of it.

## The Heroic Travels of Harriet Tubman

Although Sarah didn't have any women doctors to serve as role models growing up, she did meet one of the fiercest female trailblazers in history: Harriet Tubman. Harriet was one of the most famous conductors of the Underground Railroad and the epitome of grit and dedication.

Born into slavery around 1820 in Maryland, Harriet endured a childhood marked by cruelty and beatings at the hands of her masters. As a teenager, she suffered a head injury that plagued her with lifelong health issues, including seizures.

In 1844, she married a free Black man named John Tubman. Five years later, fearing a sale to a new master, Harriet decided to risk her life and run to the North in search of freedom. Her route? The Underground Railroad. Thanks to the help of abolitionists like Sarah's family, and a number of Quakers, who believed slavery was wrong, along with many other brave individuals, Harriet followed the North Star as she traveled by night, and eventually made it safely to Philadelphia.

It would have been understandable if Harriet had never looked back, but instead, she turned right around to help her family and total strangers embark on the dangerous journey. Over the next ten years, Harriet made nearly twenty trips to the South, leading an estimated seventy enslaved people to freedom. Each time, she risked her life, but her journeys were carefully planned, and she was never captured and never lost a passenger.

And her service didn't stop there. Harriet also served the Union Army as a nurse and even a spy during the Civil War. She was the first person to lead an armed expedition to liberate enslaved people, which guided more than 700 people in South Carolina to freedom.

When the Civil War ended, Harriet became an advocate for women's rights, especially the right to vote. A freedom fighter until the end, Harriet's legacy continues to inspire others with her strength and determination.

In 1861, the Civil War broke out, pitting the Northern and Southern states against each other. Two years later, with the war raging, President Abraham Lincoln issued the Emancipation Proclamation, which stated that "all persons held as slaves" in the rebelling Confederate states "are, and henceforward shall be free." But it would take until the end of the war in 1865, and the ratification of the Thirteenth Amendment, for slavery to be officially declared unconstitutional.

Sarah was fifteen when the war ended. Both of her parents lived to see the end of slavery, but they died soon after that freedom was declared. And by the time Sarah had turned twenty-two, her older sister Amelia had gotten married and moved away, leaving Sarah as head of the household.

The country was now in Reconstruction—an era during which the US attempted to rebuild and reunify the country. It was a time marked by rapid change and concentrated efforts to expand the rights of Black Americans. It was also the time when Sarah found her calling.

Now twenty-three, Sarah was at a train station returning from a visit to her sister when she noticed a little boy hanging feed bags around the necks of horses. Suddenly, she heard a scream. She was horrified as the child was dragged beneath a heavily loaded wagon. Sarah called out, but help was slow to arrive.

Witnessing the child's pain and being unable to assist him

was excruciating to Sarah. On the train ride home, she became determined that she would never be put in that position again if she could help it. She decided then and there that she would become a doctor.

Just months later in 1873, Sarah applied and was accepted to the newly established Syracuse University College of Medicine. In her class of seventeen students, Sarah was the only Black person, male or female, but she was not the only woman. The campus was not far from Seneca Falls, where, a quarter century earlier, the first women's rights convention had been held. (A convention, you may recall, that fellow forgotten scientist Eunice

Newton Foote attended.) One of the demands coming out of that convention was to get more women into the field of medicine.

Syracuse University heard the call. The university had been founded in 1870 by forward-thinking Methodists. The medical school was added a year later—just two years before Sarah would enroll. Fortunately, the Methodists had an open-minded approach when it came to who could be educated on their campus. From the very beginning, they allowed not just men, but also women and people of color. Sarah was a member of this new and growing modern group.

That isn't to say the road was always easy. Sarah did encounter racism during her schooling. Once, while treating a Black woman in a hospital, she was challenged about her race and gender. The patient told her, "I don't want no colored woman doctoring me." But Sarah did not let this incident deter her.

Sarah earned her MD in the spring of 1876, making her the first Black woman to do so at a coeducational institution. She then took on not one but two internships for additional medical training, which was highly unusual at the time. After completing her clinical training, she moved to Washington, DC, in 1879 and opened a private medical practice.

The Reconstruction era had removed certain obstacles that had existed before the Civil War and allowed Black Americans like Sarah to rise to positions of power and influence. Sarah was now seeing patients—who affectionally gave her the nickname "Miss Doc"—and had close friends and family living nearby, including Frederick Douglass. When Sarah's father was still alive, he had been friends with Frederick, and Sarah's sister Amelia had married Frederick's son Lewis. One of Frederick's other sons, Charles, knew Sarah and made an introduction that would once more change the course of her life.

Charles Douglass had been living in the Dominican Republic as vice-consul of the United States. While there, he made a best friend also named Charles—Charles Fraser—who he thought would be a perfect match for Sarah. Charles Fraser was a chemist—at that time another name for a pharmacist—and was planning a trip to the United States to stock up on supplies. It was quickly arranged that Charles and Sarah should meet. But when the day came, Sarah never showed. She was too busy working. Later, she wrote an apologetic letter, and she and Charles became pen pals. After several exchanges, in 1881, Charles used one of his letters to propose.

As romantic as the story might sound, Sarah wasn't so sure about Charles. She felt like they barely knew each other. Plus, Sarah was focused on building a thriving medical practice in Washington, DC, and Charles was living on an island in the Caribbean. How would they make it work?

Sarah needed some convincing. It would be the famously

persuasive Frederick Douglass who ultimately helped make up her mind, telling her, "The Dominican Republic is where you can do your best work." Sarah packed her bags.

By this point, it was the early 1880s. Reconstruction in the US was winding down and the Jim Crow era was on the rise. Life was rapidly shifting for Black Americans. They had been freed from enslavement, but the introduction of Jim Crow laws enforced the racial segregation of Blacks and whites, meaning that Black Americans were not allowed to participate freely in society and whites had considerably more privilege. The Jim Crow laws affected most aspects of daily life. Blacks and whites were kept apart in public places like restrooms, restaurants, trains, and parks. Signs stating WHITES ONLY and COLORED were posted everywhere.

Freedman and famed orator, writer, and abolitionist Frederick Douglass.

In the last quarter of the nineteenth century, the Dominican Republic was also undergoing tremendous change. Although it had won its independence from Haiti in 1844, by the early 1860s, Spain occupied the country. Dominican nationalists fought back and won their sovereignty again in 1865. Like the US during its brief Reconstruction era after the Civil War, the Dominican Republic in the 1870s and 1880s was also trying to repair and reinvent itself with railroads, technology, and telegraph lines.

The Dominican Republic had abolished slavery in 1822—more than forty years before the United States. So when Sarah and Charles arrived in 1882, they were among aspiring immigrants who were establishing themselves in the city of Puerto Plata. Charles was already friends with some of the most important political people of the period, and the couple found themselves members of the upper class of Puerto Plata society.

A look inside the pharmacy that Sarah and Charles Fraser ran in Puerto Plata.

Both Sarah and her husband were of mixed race and had relatively light skin. In the United States, the "one-drop rule" meant that a person was considered Black if they had *any* Black ancestry—and would therefore be subject to all legal and social racism and restrictions. In the Dominican Republic, however, being of mixed race meant that Sarah and Charles enjoyed more freedoms, and it gave Sarah certain privileges she never would have experienced in America.

Still, for Sarah, who had left her medical practice behind, there was one big obstacle to settling into her new life. Unlike her husband, she didn't speak Spanish. Lucky for her, she was befriended by the former president of the Dominican Republic, a forward-thinking Catholic Bishop named Padre Fernando Arturo de Meriño. He was interested in modernizing the Dominican Republic and supporting women's education, and he took Sarah under his wing and encouraged her to earn her license to practice medicine in the Dominican Republic. Not only did he tutor her in Spanish, but he also taught her the scientific terms required to pass the medical exam. Less than a year after arriving in the Dominican Republic, in the spring of 1883, Sarah took those grueling tests, in Spanish, and she passed. She was given a certificate authorizing her to treat women and children.

It was so unprecedented to have a woman practicing medicine in the Dominican Republic that the Dominican Congress actually had to pass a rule stating that Sarah could do her work. And when Congress made that exception, Dr. Sarah Loguen Fraser became the first woman licensed to practice medicine in the Dominican Republic.

However, Sarah's triumphs did not extend to all women in the Dominican Republic—at least not yet. The law stated that Sarah was the *only* woman allowed to practice medicine. It would be some time before others could follow in her footsteps, though Sarah's efforts went a long way toward ensuring that they could.

Because Sarah was restricted to treating women and children, her specialty became pediatrics and obstetrics, with a focus on labor and delivery. Giving birth was dangerous for women, which Sarah knew firsthand. A few years after receiving her medical license, Sarah gave birth to her daughter, Gregoria. However, she had such a difficult delivery that she could not have any more children. According to Gregoria, Sarah worked to ensure that other women wouldn't have to endure anything like her own childbirth experience.

For the next decade, Sarah lived and worked in Puerto Plata, treating women and children and raising Gregoria. By all accounts, as a licensed doctor trained in the latest techniques in the United States, she was highly respected.

Heartbreak struck in 1894 when Sarah's husband, Charles, suffered a stroke and died. After his death, Sarah gave up her medical practice, and for a while, she ran Charles's pharmacy. And though she'd spent a decade building a life in the Dominican Republic, Sarah and her daughter moved back to Washington, DC, to be closer to family, and to get the now-teenage Gregoria a better education.

Upon their return to the United States in 1897, however, Sarah and Gregoria faced a racial landscape that was entirely different from the one in the Dominican Republic—and from the America Sarah had left in 1882. Since she had been away, the US Supreme Court had formally legalized segregation in 1896 with *Plessy v. Ferguson*. The doctrine of "separate but equal" was now officially the law of the land. This was a time of heightened racist terrorism against Blacks, particularly in the South. Sarah, who had reaped the benefits of American higher education, now had a hard time finding high-quality schooling for her own daughter. With few good options under "separate but equal" legislation, she sent Gregoria to a boarding school in France, something only a privileged few could afford.

## Separate but Not Equal

Jim Crow laws enforced racial segregation down to telling Black passengers where they could sit on a train. Black individuals often had to sit in the back of railcars or give up their seats if a white customer came aboard and wanted it.

A man named Homer Plessy, who was part Black, challenged the unfairness of this law by sitting in a Louisiana train car that was labeled for whites only. Though Plessy's great-grandmother had been Black, the rest of his family was white, and he looked white. When he handed his ticket to the conductor in the whites-only car, he told the conductor he was one-eighth Black. The conductor asked him to leave the car. Plessy refused and was forcibly removed.

Plessy's stand against injustice that day culminated in a court case called *Plessy v. Ferguson* that the US Supreme Court heard in 1896. In the end, the court ruled against Plessy in a landmark decision. They stated that it was fair to have separate facilities for Black and white people—as long as they were "equal."

In reality, the facilities were certainly separate, but they were hardly equal. The trains, restaurants, hotels, theaters, schools, restrooms, and drinking fountains for Black communities were almost always inferior to the facilities that were provided for whites. This inequity fed into the racist attitude that Black people themselves were inferior and didn't deserve the same treatment as whites.

This widespread discrimination went on for fifty years, until, in 1954 during the civil rights movement, the Supreme Court heard another case: *Brown v. Board of Education.* A Black girl named Linda Brown wasn't allowed to attend the white elementary school near her home. Instead, Linda had to cross a railroad track and get on a bus that would take her to a different school.

Linda Brown (bottom left) with her family at a press conference ten years after the momentous *Brown v. Board of Education* decision.

This time, the Supreme Court sided with Linda and overruled *Plessy v. Ferguson*, stating that it was unfair to separate kids in schools because of their skin color. This ruling made separate facilities legally unfair and was the official end of the Jim Crow laws—but hardly the end of racism.

Sarah also struggled professionally. Even though she was multilingual and had two medical degrees and nearly two decades of experience as a private medical practitioner, she couldn't find a job. Finally, in 1908, fourteen years after the death of Charles, things seemed to be turning around. She obtained a government appointment as the resident physician at Blue Plains Industrial School for Boys in southeastern DC. However, when she arrived, they changed her job description. Instead of being a doctor, she spent her days cooking, cleaning, washing, and ironing for fourteen boys. She was a doctor, not a housekeeper! She quit.

In 1911, Sarah bought a house in DC, which she likely paid for with the earnings from Charles's pharmacy and money from the Fraser family. Gregoria and her husband, John, moved in, too, and the family lived there happily for years. Toward the end of Sarah's life, she finally did get some recognition. In 1926, Howard University, a historically Black research university, invited her to be a guest of honor at their alumni dinner. She had not attended Howard, but the university wanted to recognize her accomplishment of thirty years of practicing medicine and commend her for leading the way for Black female doctors. Sarah had helped shift people's thinking and allowed them to see that women could become doctors.

By the late 1920s, Sarah had developed kidney disease and severe memory loss. With her daughter, Gregoria, at her side, she died in 1933 at the age of eighty-three. Dr. Sarah Loguen Fraser may not have been a household name in the United States at the time of her death, but when word reached Puerto Plata, there was deep mourning. For nine days, flags were flown at half-mast, and a high Mass was held for her at the Catholic church in Puerto Plata.

In 1939, Gregoria went back to Puerto Plata and was overwhelmed by the number of people who remembered her mother.

While they cannot be sure, curators at the National Museum of African American History and Culture believe this to be a photo of Dr. Sarah Loguen Fraser carrying her medical bag.

"I spent ten months in Puerto Plata, and not a week passed that someone did not come bringing gifts of flowers, fruits, sweetmeats, the donors saying, 'Your mother operated on my daughter and she was made well.' 'I would've had consumption, but your mother taught my mother what to do.' Or, 'Your mother treated my father and took no pay. I still appreciate her kindness,' " wrote Gregoria in an unfinished biography on her mom that she titled *Miss Doc.*

It's hard to know whether Dr. Sarah Loguen Fraser viewed herself as an inspiring role model or even as a pioneering physician. But one thing is clear: The vital importance of helping Black women and children was instilled in her as a young girl providing aid on the Underground Railroad. And she further set her mind to leading in this manner when she witnessed that young boy in need of medical attention all those years before. In her journal, Sarah wrote, "To have those of my race come to me for aid, and for me to be able to give it, will be all the heaven I want."

# ELIZEBETH SMITH FRIEDMAN

AUGUST 26, 1892–OCTOBER 31, 1980

"She was like the secret weapon for the government."

—Jason Fagone, author of *The Woman Who Smashed Codes*

| CLAIM TO FAME | America's First Female Cryptanalyst |
| --- | --- |
| WHY | For her pivotal deciphering of critical coded messages, which aided the US Department of the Treasury during the Prohibition era and US military efforts in World War I and World War II |

## FIRSTS

* Worked at Riverbank Laboratories, one of the first organizations to promote the study of cryptology—the creation and breaking of coded messages
* First woman to co-lead a US military code-breaking unit

Nothing in Elizebeth Smith Friedman's early life suggested that she would go on to be a code-cracking phenom whose skills would be called upon in not one but two world wars, or that her puzzle solving would take down criminal masterminds at the height of Prohibition. What was clear from the start, though, is that Elizebeth was driven to write the rules for her own life.

Elizebeth was born in 1892 and grew up on a dairy farm in Indiana. She was the youngest of nine children and raised in the Quaker faith, a religion with a long tradition of opposing war. As a kid, she loved to read and write poetry, and her passion for knowledge only grew as she got older. Near the end of high school, Elizebeth told her parents that she wanted to attend college. But that was not typical for women at the time, and her father didn't want her to go. Nevertheless, Elizebeth was determined, and she talked her dad into giving her a loan so she could attend—the deal was that she would have to pay him back, including interest.

## What's in a Name?

You may have spotted the unique spelling of Elizebeth's name. That's no typo. Sopha Strock Smith did *not* want her daughter to go by the nickname Eliza. To avoid this, she simply changed the spelling!

In 1915, Elizebeth graduated from Hillsdale College in Michigan with a degree in English literature. After a year of teaching at a rural school, Elizebeth decided she'd had enough of the quiet life. Seeking adventure, she set off for Chicago to look for a job. But after a week of searching, Elizebeth found nothing and realized she would have to go home. On her last day in the city, she decided to visit the Newberry Library to see Shakespeare's First Folio, a collection of his plays that had been printed in 1623. The librarian noticed Elizebeth's interest, and they started talking. Elizebeth told the librarian about her job search, saying she was especially interested in literature and would like something "unusual."

As luck would have it, the librarian had a lead on a job that was quite peculiar. She told Elizebeth about a multimillionaire named Colonel George Fabyan who lived outside Chicago, in Geneva, Illinois, on a vast estate called Riverbank.

On this property, he had established Riverbank Laboratories, a company that conducted research and occasionally published scholarly works. George had an array of quirky projects—and a seemingly endless amount of money to spend on them. One project in particular seemed like a perfect fit for an English literature major like Elizebeth: finding and deciphering hidden messages in Shakespeare's plays and sonnets. While the task may appear far-fetched, there was a belief that Sir Francis Bacon, the seventeenth-century aristocrat and philosopher, authored some of Shakespeare's works. The theory went that proof of Bacon's writing could be found in secret codes embedded within the plays and poems. George was *obsessed* with finding out if this rumor was true.

The librarian introduced George and Elizebeth to each other, and George invited Elizebeth back to his estate that very night. Elizebeth didn't even have time to pack a bag! George's chauffeur drove them to the train station, and they were whisked to the Illinois countryside. The year was 1916, and twenty-three-year-old Elizebeth was about to start a job that was far more extraordinary than she had bargained for.

Elizebeth joined the group of scholars and poets who were also deciphering Bacon's alleged clues. Elizebeth had to master a code system of Bacon's own creation. The work was tedious, but Elizebeth realized that she had a knack for hunkering down and concentrating for hours. She puzzled through hundreds of pages of Shakespeare's texts for over a year, but no secret codes emerged. Intriguing as the idea was, there was no evidence that Sir Francis Bacon was the true author. However, Elizebeth had uncovered something else.

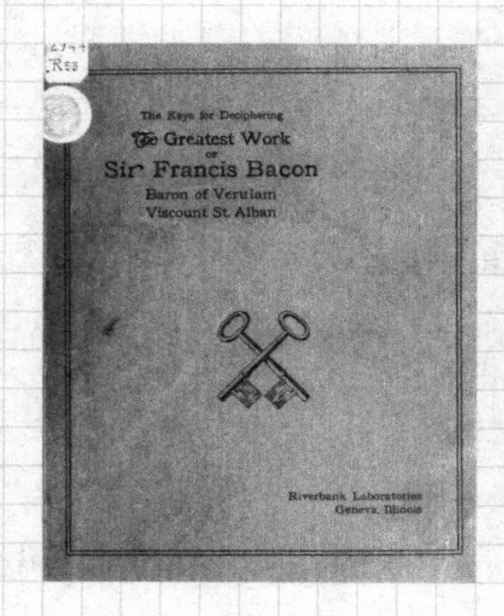

A copy of *The Keys for Deciphering the Greatest Work of Sir Francis Bacon* published from Riverbank Laboratories, as well as some of the tools scholars like Elizebeth used to create keys to investigate Bacon's writing.

It was now 1917, and the United States was about to enter World War I. The Germans and their allies were using telegram and radio technologies to send secret military information, or intelligence, to each other. The US government had figured out how to intercept these messages, but there was a huge problem: They were all encoded.

Whom could the US government turn to for help? Well, they needed individuals called cryptanalysts, who were experts in the study of making and breaking codes. While code-breakers had existed in the past, the need for them died down after the Civil War, leaving few trained cryptanalysts capable of handling all the work that was necessary at that time. So the government reached out to what seemed like an odd choice—a team of people living in rural Illinois who had been working through literary puzzles for some time. Elizebeth and her fellow Riverbank Lab scholars and poets were about to become essential for the country's wartime code-breaking efforts.

## The History of Cryptology

Cryptology is the study of codes and ciphers, and encompasses both cryptography, the turning of messages *into* codes, and cryptanalysis or code-breaking, which is *cracking* those codes.

There are numerous ways to encrypt or obscure a message so that (hopefully) only a few people can read it, and new techniques are being invented every day. Below are just a few devices that groups have used over time.

A **cipher** converts normal text into a secret message by using a formula. Such formulas range from the simple substitution of a letter with a number or symbol (called a **substitution cipher**) to more complex methods involving algorithms, which can require multiple steps to finally read the solution. The ancient Egyptians, the Roman military led by Julius Caesar, and the fictional Sherlock Holmes all used substitution ciphers in their day!

As tricky as these substitution ciphers might be, in the ninth century, Arab scholars discovered a technique called **frequency analysis**, which was an effective tool to crack the codes. The idea is to notice how often various cipher symbols appear. Symbols with the highest frequency are likely to represent a frequently used letter, such as *E* in the English language. Of course, code-*makers* found a way to throw off code-breakers using this approach. They learned to use multiple symbols to represent common letters so that the symbols would stand out less. This strategy is called a **homophonic cipher**.

During the Renaissance, between the fifteenth and the sixteenth centuries, and the Enlightenment of the seventeenth to the eighteenth century, diplomatic correspondence in Europe used a mixture of homophonic ciphers and **polyalphabetic ciphers**, which would replace frequently used letters with multiple letters but also might use those letters elsewhere to represent something entirely different in the message. (Tricky, right?)

The two world wars advanced cryptology significantly with the addition of the radio. Radio communication meant that messages could be sent over long distances and received instantly. But radio waves were not secure, so there was the risk that anyone could intercept those communications.

During World War I, the Allies were able to break almost all of the German ciphers, which were created by using codebooks and other manual methods. To make the codes more complex—and harder to break—**machine ciphers** were developed between the wars, and were heavily used in World War II. The German Enigma machine proved especially challenging for the Allies. Just when the Allies thought they had cracked Enigma's secrets, the machine would evolve to develop even more complex messages. The communications the Allies were able to interpret played a crucial role in the war's outcome.

In the 1960s and 1970s, cipher machines were replaced by computers. The codes and ciphers became more sophisticated, and the cryptanalysis techniques needed to understand them had to be even more powerful. In 1976, three cryptographers named Whitfield Diffie, Martin Hellman, and Ralph Merkle came up with the concept of **public key encryption** to secure communication in the digital age. This system, which was first made usable by Ron Rivest, Adi Shamir, and Leonard Adleman, uses pairs of keys (one public and one private) to encode sensitive information without the sender and receiver ever having to meet. In the past, the two parties had to agree ahead of time upon the encryption system they were using. Now strangers could send messages to each other! This technique revolutionized secure communication over the internet.

These days, cryptology is embedded in our digital lives. When you send an email or purchase something online, cryptology and encryption algorithms try to make sure your private information, like your credit card number and address, are protected from hackers or unauthorized businesses. Code-makers and -breakers are all around us!

Before the US government came calling, Elizebeth had been enjoying life in the countryside with a man named William Friedman. Like Elizebeth, William was working at the Riverbank Lab, though originally as an agricultural geneticist. He had been pulled into Elizebeth's project to photograph the texts. It didn't take long for the two to discover that they shared a love of Shakespeare, ciphers—and each other.

As war got underway, though, the US saw Riverbank Laboratories as a unique asset. In previous wars, the military had communicated through couriers and telegraph lines. Now, for the first time, a radio could be used to send messages over the airwaves. The problem was that anyone with an antenna could tune in. "That really increased the need for cryptography. If you can't stop the enemy from getting the message, you need to make it so that they don't understand the message," says Jennifer Wilcox, director of education at the National Cryptologic Museum. "Which meant that on the flip side, you have to be able to break those messages to understand what the enemy is doing."

War was no longer just about weapons and force. Codes, and code-breaking, were essential to overcome the enemy. With no one else to turn to, the military tapped George Fabyan's unit of poets and scholars . . . and turned them into code-breakers.

Elizebeth and William quietly got married a month after the US entered the war, and they began their married life in a new role: as leaders of George's new military code-breaking unit. Suddenly, Elizebeth was a key player in military operations, which included training military officers on how to break codes.

Until this point, code-breaking was not methodical. There were no directions to follow. It was more like working through a puzzle with pen and paper until the solution appeared. Elizebeth, William, and their team began developing code-breaking methods that could make the process more systematic and efficient. They used mathematical techniques to figure out what kinds

of encryption they were dealing with. They also experimented with comparing different coded messages to see if they could pick out common patterns. And their innovations were working. They assisted the US military in deciphering numerous communications.

In 1918, the war ended, but Elizebeth's job of code-breaking did not. She and her husband moved from Riverbank to work for the War Department in Washington, DC, in 1921. There, they were employed to help get alcohol off the streets. The year 1920 marked the start of Prohibition, which was a nationwide ban on the sale and import of alcoholic beverages. But banning alcohol didn't stop people from buying it or selling it. The Treasury Department was particularly concerned with bootleggers who were illegally selling alcohol across the United States. To evade the authorities, these mobsters had set up an ingenious system for transmitting information about the alcohol shipments using radio networks.

The Coast Guard was onto several of the mob crime rings and enlisted Elizebeth, who was at home with her first baby, to help crack the mobsters' codes. Coast Guard officials would drop packets of coded messages at her doorstep, and as soon as Elizebeth had solved one puzzle, they would bring her new ones. In her first three months on the job, she was able to decrypt two years of backlogged messages.

Elizebeth was so adept at her job that she convinced the Coast Guard to let her lead her own code-breaking unit in 1931. A role like this one in the US government was unprecedented for a woman in this era, but as she had when she first asked her dad about going to college, Elizebeth stepped up to the challenge. Her unit smashed through thousands of codes and was able to help the Coast Guard track down the Mafia alcohol smugglers. She picked up patterns in the codes and figured out who owned the ships, where they were traveling to and from, and who was meeting them. She was able to inform the federal government about the Mafia's plans and helped develop strategies to stop them.

Elizebeth became the star witness in the government's case against the Mafia in 1933. "I was called to give testimony on the messages that had been sent between these people at sea and those on shore in the smuggling operation," recalled Elizebeth. "The messages, once they were deciphered, were as plain as day."

To demonstrate how the Mafia's messaging system worked, she illustrated her technique on a blackboard. She made it look so easy that she was once accused of making it up! Newspapers and magazines, however, made Elizebeth a star. She became so famous that the government had to assign her a bodyguard to keep her safe. Even after Prohibition ended, she kept working for the Coast Guard, busting organized crime rings. In just three years, Elizebeth solved over 12,000 coded messages by hand, resulting in 650 criminal prosecutions. She even helped indict associates working for the infamous Mafia boss Al Capone.

Yet by the end of the decade, Elizebeth, now mother to a daughter and a son, had largely disappeared from the spotlight. She and her family lived a seemingly quiet life in Washington, DC. The world outside was less quiet, though. World War II was approaching. By now the US understood the importance of prioritizing code-breakers and had recruited thousands of young women right out of college to help with the endeavor. Yet Elizebeth, the trailblazing, legendary code-breaker, was not among them.

# Navajo Code Talkers

Near the end of World War I, the Germans began reliably tapping the US Army's telephone conversations, thus overhearing many of the Allied forces' plans. To intervene, the US placed Choctaw soldiers on the phone lines, where they spoke in their native language. This way, even if the Germans were able to intercept the calls, they would still have to translate the language, which was very different from their own. These brave soldiers were called the Choctaw Code Talkers.

As World War II got underway, World War I military veteran Philip Johnston knew that the government was experimenting with codes in Native American languages. As the son of a missionary, he had grown up on a Navajo reservation.

He pitched using the Navajo language, but at first the US Marine Corps was hesitant because Germany and Japan had both come to realize that the US had used the Choctaw language previously. They had even sent students to the States to study Indigenous languages in case the military deployed these tactics again. However, the Marine Corps soon learned how complicated Navajo was. Even better, the language had never been written down, making it all the more complex. In 1942, the Marines recruited and trained twenty-nine Navajo men to serve as code talkers in the Pacific. (They actually recruited thirty, but one dropped out.) The job of these brave men was to translate military terms, equipment, and locations into Navajo. From there, the Navajo soldiers communicated the messages to other Navajo men stationed across the ocean. These soldiers then translated the messages back to English for the waiting military forces.

But it wasn't always a word-for-word translation. In many instances, a Navajo term did not exist for an English term. In these cases, the code talkers began to adapt the Navajo language to fit their needs. For instance, the code talkers used the word *besh-lo*, which means "iron fish" in Navajo, to mean "submarine."

Another characteristic of the Navajo language that made it particularly useful is that as many as three Navajo words can be used to represent a single letter, such as for the letter *A*. Because there could be multiple words for each letter, full words and phrases could be better encrypted—and thus less likely to be detected using pattern analysis.

The code was complicated enough that classified information could be transmitted securely and quickly because it didn't have to be decrypted. Instead, a message merely needed to be translated by the fluent speakers working on both sides of the communication. Within minutes, a short message could be translated into Navajo, sent, and translated back into English. By the end of the war, 400 Navajo men served as code talkers.

While we have the Navajo Code Talkers to thank for numerous successful military interventions, perhaps their most famous contribution came during the seminal Battle of Iwo Jima in 1945, when the US captured the island from the Japanese. During the conflict, a team of six code talkers relayed more than 800 messages error-free. The Japanese were never able to figure out the US's plans.

President George W. Bush saluting a member of the Navajo Code Talkers during the 2001 Congressional Gold Medal celebration.

The Navajo Code Talkers played a pivotal role in securing the Allied victory in World War II. Much like the women in this book, though, they were not publicly recognized for their tremendous contributions. Immediately after the war, the military said they wanted to keep the code a secret. It would take until 1982 for then President Ronald Reagan to recognize these brave men publicly. He declared August 14, 1982, National Navajo Code Talkers Day and awarded them a Certificate of Recognition. In 1992, the US Department of Defense created an exhibit honoring these men. And in 2001, President George W. Bush presented Congressional Gold Medals, the highest US civilian award given by Congress, to the original team. Finally, their heroic efforts were seen by the world.

Rather than code-breaking, Elizebeth was focusing on another crucial mission—infiltrating the Nazi spy networks in South America. The Third Reich was pressuring South American governments to aid the Nazi Party. Accessing that information was crucial to protecting the US and its allies. And Elizebeth was on the case.

"She knew what they were gossiping to each other about. She knew what the names of their girlfriends were back in Germany," says Jason Fagone, who wrote *The Woman Who Smashed Code* about Elizebeth's life. "In addition, she got these incredibly dire and ominous messages targeting Allied ships [and] sending coordinates to Nazi U-boats so that they would be able to obliterate and murder everybody on board."

The South American spies were using classic methods like book ciphers, which use an ordinary book and a sequence of numbers to create a secret key. Fortunately, Elizebeth and William had developed a system for breaking these codes—and they didn't need the book. The Nazis, however, were using increasingly sophisticated methods for crafting their messages, including the Germans' famous Enigma machine. This innovative

A surviving Enigma machine on display in Poland.

device resembled a typewriter, with a keyboard and a set of rotors that would substitute each letter in the original message for another. The letter selected would vary according to the rotor position, which was different for every letter. There were so many possible rotor positions that the correspondence was extremely difficult to work out. But Elizebeth was even able to crack some of these messages, a feat matched by only a handful of people around the world.

TRY IT!

## Create Your Own Book Cipher

Book ciphers are a simple encryption method right at your fingertips. Let's say you want to tell your friend **MEET ME AT LUNCH** but you don't want your message falling into the wrong hands. You can use a book cipher to make your communication a little harder to read:

1. First, you and a friend agree on a specific book to use as the key.
2. For each letter in the message, find a corresponding page, line, and word in the book. The first letter of that word will be used to spell out the code.
   A. For example, if you are using *this* book, you would choose page 22, line **16**, and word 2 for the letter **M**.
   B. For **E**, you select page **63**, line **9**, word **10**.
   C. For **T**, you choose page **53**, line **19**, word **6**, and so on.

3. When you are ready to pass on your note, the first line of the encoded message would be: **22-16-2 (M)**, **63-9-10 (E)**, **63-9-10 (E)**, **53-19-6 (T)** / **MEET** / **ME** / **AT** / **LUNCH**

Ready to try your hand at decoding? Uncover this message in *Lost Women of Science:*

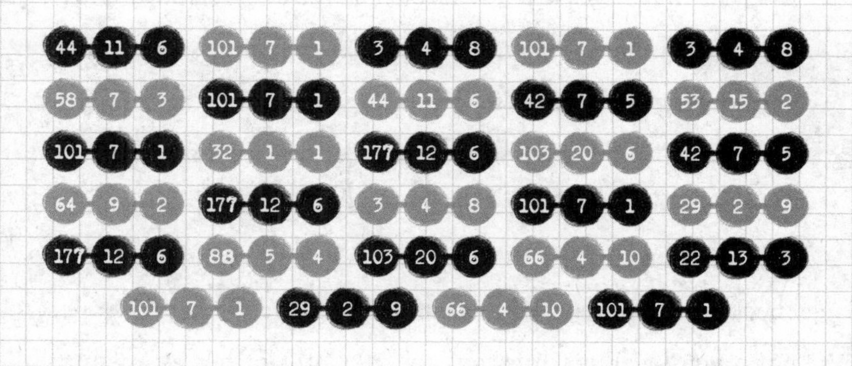


In March 1942, Elizebeth cracked a particularly important message. The code revealed that the Nazis were planning to attack the ship *Queen Mary,* which was carrying more than 8,000 US soldiers. Thanks to Elizebeth's code-breaking, the captain was warned in time and evaded the attack. Ultimately, her unit's code-breaking helped bring down the whole Nazi network in South America.

Not that the public knew. "She was a combatant in what Winston Churchill talked about as the Secret War," says Jason Fagone, "the shadow war, not a war of soldiers, but a war of languages and codes, conspiracies, radio transmitters, cipher machines." And it was all being fought under the radar. Her team decoded 4,000 messages sent on forty-eight different radio circuits. And Elizebeth never said a word. She had signed an oath in the Navy never to discuss her work, so she never did—even when the FBI and others took credit for her work. Elizebeth died in 1980 when she was eighty-eight years old and took her secret with her.

This woman who was a master of breaking secret codes and making the unseen seen was largely invisible herself. Her own family did not know the extent of her accomplishments. "My grandmother, quite frankly, never mentioned any of it to anybody," says Elizebeth's grandson Chris Atchison. "And I've checked with other cousins that knew her and they said, 'Nope, not a word.'" But while looking at photos of her as a young woman, Chris did spy hints of the person his grandmother would become. "If you look at her early portraits, you can see there's a determination in her eyes, and in her lips," he says. "She was cut from a different cloth."

Today, Elizebeth's contributions to code-breaking are on full view in an interactive display at the International Spy Museum in Washington, DC, entitled *Elizebeth Smith Friedman: The Woman All Spies Fear.* While parts of Elizebeth's life may always remain a mystery, it takes no decoding to recognize that she was a spectacular talent who changed the course of history.

# DR. CECILIA PAYNE GAPOSCHKIN

MAY 10, 1900–DECEMBER 7, 1979

> "She may have felt that she walked with giants. Now we recognize that she herself was one."
>
> —Elske v. P. Smith, *Physics Today*

| CLAIM TO FAME | The Most Prominent Woman Astronomer of All Time |
| --- | --- |
| WHY | For her revolutionary discovery of what stars are made of |

**FIRSTS**

* First person to correctly surmise that stars are primarily composed of hydrogen and helium gases
* First person to receive a PhD in astronomy from Harvard University's Radcliffe College
* First woman to chair a department at Harvard University
* First woman awarded the Henry Norris Russell Lectureship prize

Cecilia Payne possessed a curiosity that would (quite literally) reach for the stars, but she started from humble beginnings. Born in 1900, Cecilia spent her childhood in the quaint village of Wendover in England. Her father, a lawyer, died when she was four, leaving her mother to raise three children. When Cecilia was twelve, the family moved to London, and Cecilia enrolled in a church school that had a room dedicated to science on a top floor.

"I used to steal up there by myself, indeed I still do it in dreams, and sit conducting a little worship service of my own, adoring the chemical elements," she wrote later in her memoir.

As an inquisitive student, Cecilia insisted on learning advanced mathematics and German—necessary subjects if one wanted to become a scientist, but no other girl in her school had ever needed or requested those courses. Described as a shy girl, Cecilia was nonetheless insistent in her pursuits, so much so that it may have contributed to her getting kicked out of that school days before her seventeenth birthday. Stories vary, but some say Cecilia wouldn't stop entering the science lab without permission, while others allege she would pull off a variety of antics, such as putting a fake cover on her books so her teachers would think she was reading the Bible when really she was studying philosophy.

Luckily, though, Cecilia's next school, St. Paul's Girls' School, was more supportive of her ambitions. And in 1919, Cecilia won a scholarship that allowed her to attend Newnham, one of only two women's college at Cambridge University in England. At her new school, Cecilia was studying botany when she attended a lecture by astrophysicist Arthur Stanley Eddington that would change the course of her academic career and her life.

Arthur's presentation discussed his recent observations of the 1919 solar eclipse, during which he was able to prove Albert Einstein's theory that light bends when it passes near the sun was correct. "Cecilia Payne was completely transported," says author Dava Sobel, whose book *The Glass Universe* tells the story of the women who worked at the Harvard College Observatory. "She didn't sleep for three nights and she wrote down his entire lecture from memory."

Cecilia was smitten with astronomy, but her early college years were a tough time to be a female student in England. Just getting to class meant that she had to don proper attire, which meant wearing a hat and a long dress at all times, even when she was riding her bike to class. She also had to be chaperoned if she was going to any event where men were to be in attendance.

Undeterred, Cecilia started taking astronomy classes with the head of the Cavendish Laboratory, Ernest Rutherford. He opened every lecture by staring directly at her and saying, "*Ladies* and gentlemen." She was the only woman present and forced to sit in the front row as per regulations. And each time the lab director said these words, her male classmates would applaud and jeer, Cecilia wrote later in her memoir. "At every lecture I wished I could sink into the earth."

Even with Cecilia's determination to learn, she knew from the beginning that earning an actual university degree was out of the question. Women could attend the same classes and complete all of the same coursework as men, but at the end of the program, the men would receive a diploma and the women would be presented with a certificate of completion. Only the men were awarded the credentials necessary to get a job as a scientist upon graduation.

But it was 1919, and times were changing. The worldwide suffrage movement was in full swing. Just as American women were rallying for the right to vote, the women at Cambridge's Newnham College were calling for change as well. They wanted the right to earn an actual degree for the work they were doing. But when they asked for it, the men were outraged. "The male students attacked the college," says Payne-Gaposchkin biographer Marissa Moss. "They tried to batter down the women's college. They tried to put a battering ram into the gates." It would take until 1948 for Cambridge University to give diplomas to women.

Some men did go out of their way to help Cecilia, like Leslie John Comrie, known as LJ. When Cecilia discovered that Newnham College had its own observatory but it was in need of repair, LJ became her ally. He helped Cecilia fix the telescope, showed her how to use math to predict celestial movements, and explained computational astronomy.

Cecilia soon made another important acquaintance when Harvard astronomer Harlow Shapley came to deliver a lecture in London. It did not take long for Cecilia to tell Harlow that she wanted to come and work for him in America, where the suffragist movement had made inroads for women. "She had guts, she had drive, and for a shy person, really managed to tap the right people at many points in her career," says Dava Sobel.

Once Cecilia learned that there was a one-year graduate research fellowship at Harvard University earmarked for a woman, she applied for the opportunity and, to her delight, got it. There was a problem, though. She had no money to travel to America, let alone pay for living expenses when she got there. Cecilia would not be deterred. Cleverly, she entered an obscure essay contest, writing about the Greek text of one of the Gospels. Cecilia knew *nothing* about this topic, but she gambled that few people would enter the competition—and the wager paid off. No one else had submitted an entry, and Cecilia won fifty pounds, which was not life-changing wealth but was enough to jump-start her travels to the Harvard Observatory in Cambridge, Massachusetts.

Built in 1839 in what was once rural countryside, this laboratory for the night sky was a one-of-a-kind place for astronomers to view and chart the stars. The observatory's first telescope, the fifteen-inch Great Refractor, was the largest in the United States when it was first installed in 1847 and would remain so for the next twenty years. In the mid-nineteenth century, the telescope was capable of producing beautiful, never-before-seen images of the moon that were remarkably clear.

By the time Cecilia arrived in 1923, Harvard had a large collection of stunning imagery of the night sky. For a hundred years, until the advent of modern computers, these photographic images were considered state-of-the-art. The negatives were preserved on glass plates, and they were dense with information. The windowpane-sized pieces of glass held thousands, if not millions, of points of light that represented stars in our galaxy. Some were made with prisms that fanned out the starlight, creating what is known as stellar spectra. The photography allowed astronomers to literally look back in time. They could see what a star looked like ten years ago by viewing the glass plate, just as we can review old photographs today.

However, analyzing the plates was labor-intensive. That essential but tedious work fell to women. Why? Because of money. The salary of one man was equal to those of four women. So women began to fill the astronomy building. What first started as a group of astronomers' wives, daughters, and one exceptionally smart woman who worked as a housemaid soon grew into a collection of stellar women who were astronomers in their own right, even if they were undervalued. When Cecilia arrived, these "female computers" had been hard at work at Harvard for about forty years, calculating the math required to interpret astronomical observations.

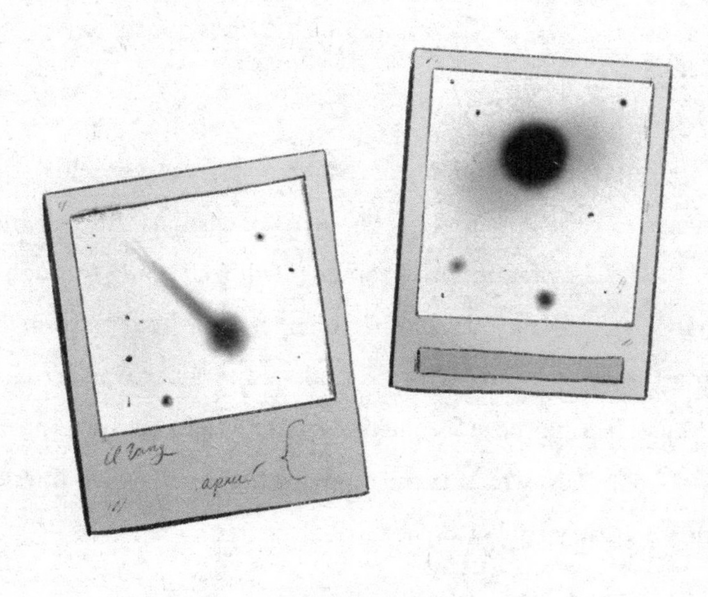

Cecilia Payne (back row, second from the left) stands with fellow "female computers" outside the Harvard Observatory in 1925.

Cecilia's fellowship meant that she was free to go where the science took her and to draw her own conclusions. At the time, the study and practice of astronomy had been going through a change as well. Scientists were shifting their focus from mapping the heavens to trying to understand what was actually going on inside the stars themselves. The field became known as astrophysics, which is the study of how the laws of physics and chemistry apply to space.

Cecilia had studied atomic physics and had at her fingertips a vast collection of stellar spectra from the Harvard Observatory. She recognized the extraordinary potential in these glass plates and was determined to crack the code of the spectra.

The popular belief at the time was that stars were made of the same commonplace elements that make up Earth, like aluminum and iron. But the more Cecilia studied the glass plates, the more she wondered whether that current thinking was, in fact, wrong. To her eyes, the hydrogen lines were far more pronounced than expected. And the second-most-abundant element she was identifying on the glass plates was helium. "And that's where she had her giant 'aha' moment," explains Dava Sobel.

Cecilia became convinced that it was helium and hydrogen—the two lightest elements—that were the primary components of the universe, not the denser elements like aluminum and iron. But this was such a radical notion that nobody could believe it. "It would be as if you said, 'The earth really is round, not flat,' " says Marissa Moss. "It just reshifts the way you look at the universe."

Cecilia's boss, Harlow Shapley, sent her findings to the leading expert in stellar composition at the time to get his opinion. Princeton University's Henry Norris Russell told Cecilia that her results must be wrong. Cecilia knew that Henry's authority could make or break a young scientist. So she made a decision she would later come to regret. Reluctantly, she agreed with

Harlow and Henry, and wrote in her thesis that her results were "almost certainly not real." However, she left her data intact, perhaps hoping her research would speak for itself.

Cecilia's work did finally earn her a PhD, although it was accredited by Harvard's female equivalent, Radcliffe College, since Harvard was an all-male university at the time. It wasn't until a few years after Cecilia received her PhD that other research came out to support her initial claim.

Using a different process, Henry, the very person who had decried Cecilia's work initially, was able to draw the same conclusions: Stars were largely composed of hydrogen and helium gases, making gas the major component in the universe. Now Henry did briefly acknowledge Cecilia's discovery, but he failed to mention that he had disputed it, and ultimately, he authored the paper that was generally accepted by the predominantly male astronomers. He received the credit.

Cecilia criticized herself for not sticking up for what she believed in. "I was to blame for not having pressed my point. I had given in to authority when I believed I was right. I note it here as a warning to the young," Cecilia continued in her memoir, "If you are sure of your facts, you should defend your position."

Getting the proper credit might have set Cecilia up to pursue bigger and better discoveries. But we'll never know, because that didn't happen. Instead, Cecilia was back to scraping by to support herself. She was used to it. From her earliest days as a student, she substituted grit and elbow grease for the funding she lacked. "She was too poor to buy books. She was copying textbooks by hand," says biographer Marissa Moss. After Cecilia earned her PhD in 1925, her fellowship funds stopped. She needed a job and was hired back at Harvard by Harlow Shapley under the title of technical assistant.

"I was paid so little," she wrote. "I was ashamed to admit it to my relations in England." Even though she had to sell her jewelry and her violin to get through the first month, she focused on the opportunity the job did provide: access. "I had the run of the Harvard plates. I could use the Harvard telescopes, and I had the library at my fingertips," she went on to say in her memoir.

Cecilia taught many astronomy lectures, although she wasn't listed in the course catalog. Her boss tried and failed to get her promoted, as Harvard wasn't ready to add women to the all-male faculty. Then, when Cecilia was around thirty-three, tragedy struck. In the span of one year, she lost two friends, both to boating accidents. To take her mind off her loss and rethink her life, she traveled to Europe. In Germany, she met a thirty-five-year-old Russian astronomer named Sergei Gaposchkin. At the time, Hitler was taking power, and Sergei had lost his job. He was suspected of being a Russian spy by the Germans, and a German spy by the Russians. He was stuck without a country to call home.

Cecilia turned to her mentor, Harlow Shapley, who helped Sergei get a visa and gave him a job at the Harvard Observatory working directly with Cecilia. After just a few months, they fell in love, went off to New York, and got married. Their colleagues disapproved of the match, in part for the silliest of reasons, because she was taller than him.

Still, they formed quite a pair and worked closely at the Harvard Observatory, dividing up the universe between them. Much of their research was on variable stars, which change in brightness over time and in regular cycles. The couple worked on parallel tracks, but with a clear common interest in all things celestial, while their three children ran through the maze that was the observatory complex, playing games like hide-and-seek.

Cecilia relished being a wife and mother. "I had once pictured myself as a rebel against the feminine role. But in this, I was wrong. My rebellion was against being thought, and treated, as inferior," Cecilia wrote. She also clarified that she never saw herself as a female astronomer. Of the title scientist or scholar, she noted, "Neither of these words has a gender."

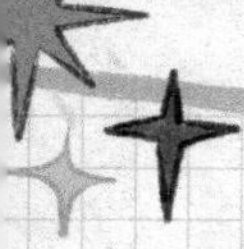

# Supernova Astronomers

Today, many female astronomers and astrophysicists continue to break new ground and contribute to our knowledge of the stars and the universe. **Sara Seager**, a professor at the Massachusetts Institute of Technology, and **Lisa Kaltenegger**, the founding director of the Carl Sagan Institute at Cornell University, are at the forefront of searching for exoplanets, or planets beyond our solar system, that contain atmospheres and signs of life. Astrophysicist and professor **Natalie Batalha** of the University of California, Santa Cruz, was the scientific lead for NASA's Kepler mission, which led to the discovery of thousands of exoplanets and exoplanet candidates that are yet to be confirmed. Astrophysicists **Priyamvada Natarajan** at Yale and **Feryal Özel** at the University of Arizona have both worked with the Event Horizon Telescope Collaboration and paved the way to groundbreaking discoveries about black holes. **Katie Bouman**, an associate professor at California Institute of Technology and a scientist also on the Event Horizon team, helped create the algorithm and sensor design that allowed for the first photo of a black hole. **Ewine van Dishoeck**, a professor at Leiden University, is delving into molecular astrophysics and the chemistry of space, in addition to searching for evidence of water and the water cycle in space, and Harvard professor and Smithsonian Institution research associate **Alyssa A. Goodman** is combining astronomy, data visualization, and digital scholarship to glean more information on the interstellar medium and star formation. And that's just a few of today's female scientists studying stars to help us better understand the universe.

After a long history of preventing women from joining the faculty, Harvard finally made Cecilia a full professor in 1956. She was the first woman to be elevated to that position from within the university. The promotion came with a sizable—and long-overdue—pay increase, and a few months later, Cecilia Payne-Gaposchkin became the first woman to chair a department at Harvard. She was fifty-six years old and held the title for four years, until 1960.

By then, Cecilia had written or cowritten 9 books and 351 papers and had made literally millions of observations of stars. But she is best known for the PhD dissertation she wrote back in her early twenties in which she cracked the code of the secrets within stars. Four decades later, her research still holds up. In 1962, astronomer Otto Struve pronounced Cecilia's PhD doctoral dissertation "the most brilliant thesis ever written in astronomy."

In 1976, three years before her death, Cecilia became the first woman awarded the Henry Norris Russell Lectureship prize. The award was indeed named for the man who, decades earlier, had doubted her discovery.

# The Stars Aligned: A Timeline of Astronomical Exploration and Discovery

## 600 BCE—0 CE

* The ancient Greeks develop models of the universe. They view the stars as being fixed in space.

## 2ND CENTURY CE

* Egyptian astronomer Ptolemy compiles the *Almagest*, a text that describes over 1,000 stars and creates a model of the universe.

## 10TH CENTURY

* Persian astronomer al-Sufi writes *The Book of Fixed Stars,* based on Ptolemy's earlier work.

## 16TH CENTURY

* Polish astronomer Copernicus is the first to suggest that Earth and the other planets orbit the sun.

## 17TH CENTURY

* Italian astronomer Galileo Galilei uses a telescope to see mountains on the moon, Jupiter's moons, Saturn's rings, and the phases of Venus.

* German astronomer Johannes Kepler discovers the laws of planetary motion.

* British physicist Sir Isaac Newton (relative of Eunice Newton Foote!) and others discover spectroscopy, the study and interaction of light and matter.

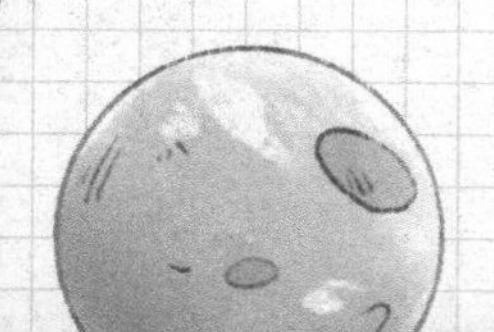

## 18TH CENTURY

* German-British sibling astronomers William and Caroline Herschel made many pioneering discoveries together. While observing the star Castor, William discovered binary stars, or pairs of stars that orbit a common center mass. And Caroline was the first woman to observe a comet, among other findings.

* French astronomer Charles Messier publishes his *Catalogue des Nébuleuses et des Amas d'Étoiles,* also known as the Messier catalog, increasing knowledge about star clusters and nebulae.

## 19TH CENTURY

* German astronomer Friedrich Bessel uses a concept called stellar parallax to determine the distances to stars, which helps people understand the universe's vastness.

* Astronomers develop the technique of capturing telegraphic images of stars on glass plates. These images allow them to travel back in time to see how stars have changed over time.

## 20TH CENTURY

* British-born American astronomer Cecilia Payne-Gaposchkin discovers that stars are made of hydrogen and helium.

* German physicists Hans Bethe and Carl Friedrich von Weizsäcker and British astrophysicist Arthur Stanley Eddington contribute to the discovery that nuclear fusion is the process powering stars.

* NASA launches the Hubble Space Telescope, allowing for the study of stars in distant galaxies, exoplanets, and the early universe. Katherine Johnson, Dorothy Vaughan, and Mary Jackson calculate early spaceflight trajectories, launch windows, and return paths.

## 21ST CENTURY

* NASA launches the James Webb Space Telescope, and astronomers are able to look further into the universe's past than ever before.

Near the end of her memoir, Cecilia offers up this advice to young women: "Do not undertake a scientific career in quest of fame or money. There are easier and better ways to reach them. Undertake it only if nothing else will satisfy you for nothing else is probably what you will receive. Your reward will be the widening of the horizon as you climb, and if you achieve that reward, you will ask no other."

# DR. DOROTHY ANDERSEN

"Luck favors the prepared mind. The prepared mind is Dorothy. She's brilliant and she put this puzzle together.

—Brian O'Sullivan, professor of pediatrics at Dartmouth Geisel School of Medicine and codirector of the Pediatric Cystic Fibrosis Center"

| **CLAIM TO FAME** | The Puzzle-Solving Pathologist in the Basement |
|---|---|
| **WHY** | For becoming the first person to identify the disease cystic fibrosis |

**FIRSTS**

* First physician to diagnose cystic fibrosis
* First physician to create a diagnostic test for cystic fibrosis and a treatment plan
* First physician to identify cystic fibrosis as a genetic disorder

Dorothy Hansine Andersen always had a knack for doing things her own way. The daughter of Danish immigrants, Dorothy was born in 1901 in Asheville, North Carolina. Dorothy had to learn to be independent from an early age. At five years old, she lost her father to nephritis, an autoimmune disease that causes kidney inflammation. And when Dorothy was nineteen, her mother, who had struggled with her health for years, died as well.

MOUNT HOLYOKE COLLEGE

After high school, the newly orphaned Dorothy went on to attend Mount Holyoke, an all-women's college in Massachusetts. Dorothy studied chemistry and zoology and earned a reputation as a young woman who enthusiastically shared her passions with others. "Dorothy was a self-appointed saleswoman for outdoor exercise and the wonders of the natural world, forever trying to involve the intellectuals around her in her strenuous pursuits," wrote biographer Libby Machol in an unpublished book on Dorothy. "One winter morning, she persuaded [her friend] Edith to join her for an early breakfast of bacon and eggs. *In the snow.* All the while she was building the fire and cooking the meal, she talked about the joys of fresh morning air. In the spring, she organized expeditions to collect frogs and skunk cabbages for the lab."

Dorothy graduated from Mount Holyoke in 1922 and turned her passion to medicine, earning a medical degree from Johns Hopkins University in Baltimore in 1926. She was one of only five women in her class. By now, though, Dorothy was used to being an exception to the rule.

## Sew Thrifty

Part of Dorothy's brilliance was her ingenuity. As a young doctor with a love of the outdoors, she found herself yearning for a sleeping bag she'd seen at Abercrombie & Fitch. But there was no way she could afford the steep price. Instead of being discouraged, Dorothy decided to construct her own sleeping bag. She found fabric, filled it with down, and sewed the creation by hand. She even tied the knots securely with the one-handed technique she had learned in medical school to stitch up patients.

Fortunately, she was not the only one going against the grain. Dorothy's lifeline and mentor was Dr. Florence R. Sabin, who also pushed the boundaries of what women were thought to be able to accomplish. In 1917, Florence became the first woman to work as a full professor at Johns Hopkins. Even with Florence's guidance, though, the male-dominated world of medicine wasn't easy to penetrate. After medical school, Dorothy applied for a surgical residency program, but despite her medical degree, she was rejected simply because she was a woman. Many hospitals hesitated to hire women as doctors. In those days, male and female patients alike reported that they had more confidence in male physicians.

If female doctors did get the opportunity to work, they were usually limited to gynecology, the care of women's reproductive organs; psychiatry, the study of mental illness; or pathology, the study of diseases. Pathology was often open to female doctors mainly because their patients couldn't complain—they were already dead!

And pathology ended up being where Dorothy was given an opportunity. In 1929, she was offered a position at Babies Hospital (known today as Children's Hospital) at the Columbia Presbyterian Medical Center in New York City. She began as an assistant in the pathology department, but by 1935, she had earned her Doctor of Medical Science. That same year, she also accepted a new role as a pathologist in the hospital. It was in this position that she would make one of the biggest medical breakthroughs of the twentieth century.

At the hospital, Dorothy's job was to perform autopsies, or surgical examinations, on children who had died to see how a disease had impacted the body's heart in particular. Autopsies can help doctors understand a lot about how our bodies work. They can reveal, among other things, how a disease has progressed over time, which organs might be affected during an illness, and whether any medical treatments show promise of helping to combat a disease. Ultimately, one hope is that autopsies can allow us to get better at treating patients. One day in 1935, Dorothy was examining the body of a three-year-old child. The prior year, the young girl had arrived at the hospital with a bloated belly, skinny limbs, and persistent diarrhea. She was diagnosed with celiac disease, which is an immune reaction to wheat and rye that inflames the intestine.

During the autopsy, though, Dorothy noticed some unusual

things. The lungs were plugged with a thick, sticky green mucus. The pancreas, a gland that produces digestive juices and hormones, had an unusual texture. "When she tried to cut it," Bijal Trivedi, author of *Breath from Salt,* a book about cystic fibrosis, explains, "she heard a scraping sound as if she were cutting through grit or sand."

Dorothy knew that celiac disease causes the walls of the intestines to become inflamed. But there was no reason why the lungs and the pancreas should be clogged. The mucus Dorothy observed was gumming up the lungs like molasses in a straw. Dorothy had come across a disease no one had named.

After the strange autopsy, Dorothy went back to review twenty cases from her own hospital and wrote letters to other hospitals, too, in search of a pattern. "She must have spent hours in the library finding these articles and then writing to other physicians and seeing if she could get their slides from autopsies to be able to compare to what she was seeing," says Brian O'Sullivan, a pediatric pulmonologist who teaches at the Geisel School of Medicine at Dartmouth College. Pulmonologists specialize in breathing and lung problems.

In 1938, after analyzing hundreds of autopsy slides, Dorothy published a fifty-page research paper called "Cystic Fibrosis of the Pancreas and Its Relation to Celiac Diseases." In today's world, a full team of scientists would likely have contributed to such an advanced, in-depth document. But Dorothy was the study's sole author.

Having combed through the numerous cases, Dorothy was able to detail in her paper the symptoms of this new disease. Patients often presented with thick mucus and damage to the lungs and digestive system. "She spelled out that she was identifying a completely different disease, and it was the first time it was recognized as a separate entity," says Brian O'Sullivan. Dorothy's paper marked the first time the disease was given the name cystic fibrosis, or CF.

## What Is Cystic Fibrosis?

Although cystic fibrosis (CF) is classified as a rare disease, an estimated 70,000 to 100,000 people around the world suffer from it, and about 1,000 patients are diagnosed in the United States each year.

CF primarily affects the lungs and the digestive system, but it gets its name from the fibrous tissue and cysts that grow in the pancreas. The disease is also characterized by the production of sticky mucus in the body. Thin linings of mucus in the human body actually do an important job. Mucus acts as a lubricant and keeps dust, pollen, and other allergens out of our lungs; it keeps our bodies moist; and it helps protect us from disease. In CF patients, though, the mucus becomes thick and builds up in the lungs and digestive system, leading to chronic lung infections and difficulty in digesting food. People with CF have a persistent cough and difficulty breathing. Many don't gain weight and fail to grow, despite having a good appetite.

Dorothy Andersen's foundational research into CF paved the way for future discoveries. In the decades since she died, the prognosis for CF patients has continued to improve. In the 1970s and 1980s, CF patients rarely lived past their teens. But these days, all newborns are screened for CF usually within a day or two after they are born—and some are still tested with an exam that Dorothy created years ago! Early diagnosis has led to longer life expectancy for CF patients, which today is around fifty years. There is no cure yet for the disease, although there are effective treatments. Just a few years ago, the FDA approved a breakthrough therapy called Trikafta™ that is giving patients and their families hope for both improved prognoses and quality of life. And it all leads back to Dorothy's tireless work in hospital basements.

Dorothy's paper noted that CF was congenital, meaning babies were born with it. She also noted that CF was common in siblings, which led her to believe that it was passed down from their parents via a mutated gene. Science would later confirm that she was right. And in the 1980s, scientists would even be able to identify which single gene mutation causes CF. If both parents pass on the gene mutation, the child will inherit the disease.

Eventually, Dorothy moved her work from the laboratory to the clinic, where she could help living patients. In late 1938, the same year she identified the disease, Dorothy created the first CF diagnostic test. A diagnostic test is a way to check off various symptoms to see if a patient may have a particular disease. In this case, Dorothy's test was to check for pancreatic enzymes, substances that break down food. In healthy patients, enzymes appear in the intestines. But in CF patients, those enzymes don't reach the intestines. So Dorothy needed a tool to help her look.

There was one issue: the tool didn't exist yet. But that didn't stop Dorothy. She tapped into the same DIY ingenuity that had once inspired her to sew a sleeping bag and build a cabin on her farm in New Jersey, and she invented a tube that could be inserted down a patient's throat, into the intestine, to collect a sample. After much trial and error, she constructed an instrument that she deemed "highly satisfactory" out of thin medical tubing with a dumbbell-shaped metal tip attached by silk thread. Dorothy's colleagues were divided in their reactions to her unusual, hands-on approach to problem-solving. Admirers praised her originality and daring. But her detractors were outraged. They considered her undignified and believed she did a disservice to her profession, and particularly to women doctors, by lowering herself to work with her hands "like a common laborer."

Dorothy was unfazed, though, and continued her research. During a 1948 heat wave, she discovered another clue to the disease when a spate of sick, dehydrated children who had CF were admitted to the hospital. Dorothy noticed that these young patients were not reabsorbing the fluids that were being administered to them through an IV. Dorothy knew that CF affected the body's glands, but now she was zeroing in on the sweat glands specifically. "She had a sixth sense about this disease that enabled her to pick out important clues," says Scott Baird, a pediatric critical care doctor at Columbia University Medical Center. It turns out the mutation in the gene responsible for CF also affects how water moves in and out of the body's cells. The result is not only the thick mucus in the lungs, but extra-salty sweat.

Dorothy wasn't the only person to observe this phenomenon.

In medieval Europe, a curse that became folklore stated: "Woe is the child who tastes salty from a kiss on the brow, for he is cursed and soon must die." Even into the 1980s, discerning doctors found they could learn a lot by kissing a baby and tasting salt.

Dorothy and another doctor named Walter Kessler wrote a paper about the connection between salty sweat and CF that was published in 1951. That link led to a new diagnostic tool—the sweat test.

## Sweat It Out

Having a tube stuck down your throat may be an effective way to be tested for CF, but it isn't the most comfortable. Luckily, Dorothy and Walter figured out a way to make the process easier on patients. With their new diagnostic tool, doctors would apply a chemical called pilocarpine on the skin and then place a small electrode on top of the area to induce sweating. After a few minutes, the doctors could collect the sweat sample and send it to the lab to examine the results.

The test measures the amount of chlorine in sweat. CF patients usually have higher amounts of chlorine than healthy patients. The results can be found reliably, painlessly, and without the patient ever having to say "Ahh!"

To this day, Dorothy and Walter's diagnostic test remains a key tool in identifying CF. However, in 1953, the credit for the sweat test went to a doctor named Paul di Sant' Agnese, who was Dorothy Andersen's mentee. Paul worked closely with Dorothy, and the two were very involved in the study of CF and the development of the sweat test, writing many papers on the subject. The problem is Paul got the public kudos, while Dorothy was forgotten.

It is unclear how Dorothy felt about not receiving proper credit for her work. But in those days, it was common for men to take the credit for women's work. So much so that today there is actually a term for it: the Matilda Effect.

## The Matilda Effect

Thirty-eight years after her death, Dorothy was inducted into the National Women's Hall of Fame for her contributions to medical science. What took so long? Some say it's the Matilda Effect, a phenomenon that explains how many men have gotten credit for women's work because of gender bias.

Margaret Rossiter, a historian of science, coined the term in 1993. Rossiter named the concept after Matilda Joslyn Gage, a suffragist and abolitionist who wrote a pamphlet in 1870 called

"Woman as Inventor." In the pamphlet, Gage wrote, "Although woman's scientific education has been grossly neglected, some of the most important inventions of the world are due to her." The Matilda Effect gives a name to the fact that women in science have been overlooked because many of their discoveries and breakthroughs were instead attributed to men.

When women are not credited for their accomplishments, it has lasting effects. If textbooks and internet searches ignore the numerous contributions of women and other underrepresented voices, it's not just wrong, it also perpetuates the "great man theory" of history. This theory explains how a single person, most often a white man, is credited for a historic breakthrough, when in fact teams of people—including many women—propel science by sharing new ideas.

Unfortunately, gender bias in science and academia is a form of discrimination that still exists today. Fewer women than men hold senior academic positions, and fewer women than men are recognized with top awards. But thanks to a concerted effort toward more inclusive hiring and promoting practices, things are slowly improving. Getting more girls and women involved in STEM (science, technology, engineering, mathematics) can play a vital role in challenging this inequity in the scientific community. As more women enter STEM jobs, they can serve as mentors and role models for the next generation.

Word got out about Dorothy's dedication to the disease, and she became known as a specialist. When Dorothy first started practicing medicine, it was rare for patients with CF to live past the age of five. Desperate parents began bringing their kids to Dorothy from all over the country. "She felt the suffering of others, and she did her best to try to minimize that whenever possible," says Scott Baird.

During the 1940s and 1950s, Dorothy was driven to share her findings about CF. She went on lecture circuits, giving talks up and down the East Coast, and even spoke at medical schools and hospitals while she was on a European vacation one summer about the importance of teaching. "She [said], 'Doctors aren't as important as teachers. The most they can do is keep people alive. Teachers train character: the heart and the mind,' " according to Libby Machol, a biographer who documented facets of Dorothy's life. "Years later, when young doctors from all over the world came to study in her laboratory, she would urge them to teach what they had learned to their colleagues when they went home."

In 1963, Dorothy died from lung cancer likely caused by smoking. She was buried in Chicago alongside her parents. The doctor, teacher, and rugged individualist never got married or had children, but she was beloved by patients, colleagues, and friends alike.

And during her lifetime, she blazed a trail, publishing more than eighty papers and bringing awareness and relief to many CF patients and their families. During her career, Dorothy was also recognized with several awards, including the Elizabeth Blackwell Award, which is given to women who have made significant strides in medicine.

Dorothy never set out to identify a disease. But she kept an open mind, applied herself to her work, and then had the courage to confront the medical community when she found that their prior conclusions were wrong. Her work pioneered almost a century of science and discovery in a time when men dominated the field of medicine. And she did it with remarkable generosity of spirit.

# KLÁRA DÁN VON NEUMANN

AUGUST 18, 1911–NOVEMBER 10, 1963

"I don't think she ever realized how remarkably intelligent she was."

—Marina von Neumann Whitman,
Klára Dán von Neumann's stepdaughter

| CLAIM TO FAME | An Early Modern-Day Computer Programmer |
| --- | --- |
| WHY | For writing the first modern-style program code ever executed on an electronic computer |

## FIRSTS

* The first to execute a modern-style code on a computer
* One of the early operators of the first electronic programmable computer
* Implemented the first computerized Monte Carlo simulation, to predict the viability of atomic bomb designs

Klára Dán von Neumann—Klári to her family and friends—was born in Budapest, Hungary, in 1911 into a well-to-do Jewish family. Klári was highly educated but didn't show any interest in science or mathematics in her early school years. Her kindness, more than her intelligence, was a defining characteristic. "She was so very open-hearted," says Agi Antal, a Hungarian translator who read Klári's letters, "nice to everybody who she loved."

By the time Klári was ten, she had lived through the First World War and a violent Communist revolution in Hungary. But despite these hardships, she continued to thrive, developing a deep love of ice-skating. At age fourteen, she even became a national figure skating champion. Her large extended family also remained tight-knit and lived together in several flats in an extravagant villa in Budapest.

Her parents were very social and hosted merry gatherings known as *mulatság,* where artists, intellectuals, and playwrights would share ideas over meals. It is likely these ideas sparked Klári's young mind as she grew to be a keen observer of the world, even as a teenager. "I [am] a tiny little speck, an insignificant insect just chirping around to see where the most fun could be had," Klári wrote in her unpublished memoir, *The Grasshopper in the Very Tall Grass.* While Klári recognized that she was but one person in a large world, it is clear she held big dreams for her future: "I don't know what is written in the book of my destiny, but I don't believe that I will ever have a normal life," she continued in her memoir. "Do I feel this way just because I'm young?"

An illustration of the diary kept by Klári Dán von Neumann.

Klári was always on the hunt for new and exciting horizons. For a woman in the 1930s, though, freedom could be hard to come by. Marriage was one way that a woman could try to gain a foothold in a new life. It was at one of her parents' intellectual parties that Klári met her first husband: a gambler named Ferenc Engle. The couple often traveled from casino to casino, and one night in Monte Carlo, the quick-witted, fun-loving Klári met another man who turned out to be a much better match. John von Neumann was a Hungarian mathematician-physicist-engineer with a genius for synthesizing and analyzing complex things. His friends and family called him Johnny. The two spent the night chatting and got along immediately.

Despite their clear affection for each other, Klári and Johnny went their respective ways that night. But several years later, in 1937, the two friends met again. By this time, Klári had divorced Engle and remarried a young banker, and Johnny had separated from his first wife. Johnny and Klári's connection was inescapable. Soon after they reunited, Klári divorced her second husband, and in 1938, a few months after Klári's twenty-seventh birthday, Klári and Johnny married. Klári moved from Budapest to the United States, where Johnny worked as a math professor at the Institute for Advanced Study in Princeton, New Jersey. The couple enjoyed married life, but World War II was imminent, and their Jewish families were still in Budapest, where their lives were under constant threat. Klári was able to return to Europe and get her family out just before Germany invaded Poland in 1939.

Safely back in Princeton, with Johnny absorbed in his work, Klári struggled to fit into suburban life. And she wasn't the only one. Her father, who had been a successful businessman in Hungary, also found the move to the United States difficult. Europe was engulfed in war, and members of his family continued to face extreme danger. Unsure of his role in this new country and unable to grapple with the mounting pressures, Klári's father died unexpectedly in 1939. Klári, who was close to her father, was devastated.

Facing grief and loneliness, Klári was desperate to find

an outlet for her intellect and energy. In one letter she wrote, "Princeton sees me as pretty enough, but otherwise . . . an average woman . . . from whom they cannot expect anything special, and the best would be if I gave birth to a kid. Otherwise, I would be very bored there." Klári eventually became pregnant but suffered a miscarriage. She and Johnny never had any children.

On December 7, 1941, the Japanese bombed Pearl Harbor and America entered the war. As men enlisted to serve their country, new employment opportunities opened for women. There was huge demand for improved weaponry to help with the war effort, and developing these weapons required a high level of math skills. Klári got a job at Princeton as the head of the Statistical Computing Group, and later at the Office of Population Research. Klári, who held only a high school degree, took a calculus class. It was clear that she had a knack for numbers.

In those days, a "computer" was a person—often a woman—who performed difficult, often tedious mathematical tasks. The work was feverish, and the computers had to perform under intense pressure. But Klári was up to the challenge. "At long last my old dream has been fulfilled, I am so busy that the twenty-four hours of the day could be doubled and I still would not be able to do everything I would like to," she wrote.

While Klári was in Princeton, Johnny was crisscrossing the country searching for the latest technology to help the war effort. Eventually, he settled at the Manhattan Project at Los Alamos in New Mexico, where he put his mathematical skills to use developing bombs, specifically the atomic bomb. On August 6, 1945, the United States dropped the first nuclear weapon over the city of Hiroshima, Japan, instantly killing about 80,000 people, most of them civilians. Three days later, the US dropped a second atomic bomb over Nagasaki, instantly killing another 40,000. Klári did not record much about her husband's involvement in developing these weapons, but she did write in her memoir that she believed the atomic bomb was "the origin of all the suicidal problems of the world today."

The scientists at Los Alamos were divided on the devastation their creations had caused and what nuclear weapons meant for the future of humanity. Many scientists left Los Alamos soon after the bombs' detonations, but Johnny frequently returned to work on what would ultimately become the hydrogen bomb. And to develop this technology, the scientists needed both more and quicker calculations. What they needed was *electronic* computer calculations.

During the war, the United States and the Soviet Union had been allies, but soon after, their relationship turned hostile. And as tensions escalated, the only reassurance the US had that an attack by the Soviet Union wasn't imminent was that the US was the only country with nuclear weapons—but that was bound to change at any moment. The Soviets were inching closer to perfecting nuclear weapons of their own.

John von Neumann's badge granting him access to the Los Alamos Laboratory.

In 1945, Klári received a telegram from Johnny asking her to move from Princeton to Los Alamos, saying, "Bring riding and skating things if possible, opportunities very good." She moved immediately. In New Mexico, there was a frozen pond where Klári could ice-skate, brilliant scientists who threw parties (much like her family's *mulatság*s of years past), and fellow Hungarians who had stories and memories of home. Klári not only reunited with her husband in Los Alamos, but they were also now working together as colleagues on a top-secret project: the most powerful bomb ever developed.

Determined to find a way to speed up the calculations needed to make the weapon, Johnny focused on finding machines to do the work instead of people. "He wanted to build a fast, electronic, completely automatic 'all purpose' computing machine which could answer as many questions as there were people who could think of asking them," Klári wrote in her memoir.

The world's first computers looked like large, intricate tape dispensers and could perform mathematical operations by reading symbols on the tape as it moved from left to right through the machines. These early computers were mechanical and relied on moving parts—gears or electromechanical switches to store and manipulate numbers.

In 1946, one of the world's first electronic computer, called the Electronic Numerical Integrator and Computer, or ENIAC for short, was developed at the University of Pennsylvania. The ENIAC took programmability to the next level. Now information could be transmitted at electronic speed, which is practically the speed of light.

Female computers Ruth Teitelbaum (crouching) and Marlyn Meltzer (standing) operating the early ENIAC machine.

But the ENIAC had limitations. It was difficult to operate, couldn't store much data, and depended on humans to plug in a complicated array of cables. Johnny wanted to help solve ENIAC's shortcomings. Luckily, he had Klári back by his side to help. "The ideal subject was right there within easy reach—namely me," Klári wrote in her memoir. "I became Johnny's experimental rabbit."

And Klári wasn't the only woman on the project. In fact, ENIAC's first operators were a team of women, who became known as the ENIAC Six.

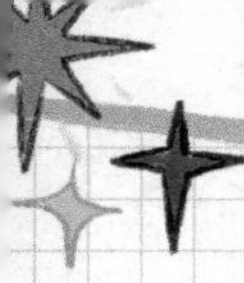

## The ENIAC Six

To get the world's first computer off the ground, it took a little girl power. The computer operators known as the ENIAC Six were Betty Snyder Holberton, Frances Bilas Spence, Ruth Teitelbaum, Kathleen "Kay" McNulty, Marlyn Meltzer, and Jean Jennings Bartik, each an accomplished mathematician in her own right.

Launching the first computer seems like an important job, so how did women finally beat out the men to get the gig? First, women had already been doing the computational labor to program the machines, so it was a natural extension of their work. Second, it wasn't considered a very difficult job. Computer programming itself was undervalued simply because engineers didn't understand the effort and expertise that went into it. The fact that the field of programming was overlooked is likely why Klári and the women in the ENIAC Six were given the chance to do it.

In reality, computer programming was—and still can be—an enormously complex job, and the women who pioneered it were severely underestimated. The ENIAC Six had high mathematical proficiency. They also had patience. Setting up computations was incredibly tedious, and the ENIAC Six had to rewire the machine by hand for every new program. Because they had no software as we understand it today, the ENIAC Six used a system of switches, cables, and punch cards. When the computer malfunctioned, they had to examine the circuitry of 18,000 vacuum tubes to find the one bad tube. (In fact, the term *debugging* was popularized when a moth got stuck in an early computer at Harvard University and had to be picked out!)

The women also had to document their work to create instructions so others could set up computations for themselves. Klári and the team understood early that if they could program instructions into the memory of the computer, the machine would know what to do from the start. That meant that if they needed to make any changes to the program, all they would have to do would be to turn the switches, instead of starting from the beginning.

And the persistence of the Six paid off. By 1948, the ENIAC had been reconfigured into an electronic computer that could be easily programmed and could store vast amounts of information. The computer could give responses as fast as instructions were loaded onto it. The ENIAC now had a high processing speed without the painstaking process of patching and rewiring cables for every new problem—and it was all thanks to the sisterhood of STEM!

The female programmers were paid less than their male counterparts, and they were assigned a task no one had ever done: to tell a machine—which can do a whole lot—exactly what to do. Klári was essentially a translator, taking a mathematician's plan of action and turning it into a language of electronic signals that a computer could understand. "I learned how to translate algebraic equations into numerical forms, which in turn, then, have to be put into machine language in the order in which the machine was to calculate it," she wrote. She compared it to solving an intricate jigsaw puzzle. If she could figure out the right pieces and put them in the correct order, maybe the computer would understand.

So how did she do it? With a pen and paper on a flowchart. Johnny created a massive document of mathematical operations, variables, and arrows pointing up and down. To the untrained eye, it looked like gibberish. But to Klári, it made perfect sense. Back then, machines weren't yet fluent in human languages, so she had to become fluent in theirs.

Klári's original role as a coder was to write down numbers that corresponded to specific instructions for the computer. These codes could tell the machine to perform several tasks on its own, for example to repeat blocks of code, as opposed to needing a human to prompt each item. Today, we call these coded instructions the modern code paradigm, and Klári was one of the first programmers to use it.

Now that the ENIAC was up and running, scientists working at Los Alamos wanted it to figure out what happens during nuclear fission, a reaction in which the nucleus of an atom splits into two or more smaller nuclei and in doing so releases a lot of energy. And specifically, they wanted to see what happened when they created stronger nuclear bombs. If the scientists could use the computer to predict how a bomb would react under specific conditions, it would save them a lot of time and money because they would no longer have to build and test the bombs themselves—saving them a lot of physical resources.

The atomic bomb was powered by a process called fission. When a neutron collides with an atomic nucleus, it breaks it apart, creating a chain reaction that results in an explosion. But with all those tiny particles of matter moving around, fission can be unpredictable. So much so that no single math equation could account for all the possible outcomes.

Scientists theorized that the ENIAC could perform a simulation called the Monte Carlo method to calculate all of these possible outcomes—if only someone could figure out *how* to tell the ENIAC to do it.

## Playing the Odds: The Monte Carlo Method

The first time Klári met John von Neumann, he was playing roulette, a casino game where a ball is placed on a spinning wheel, and players

try to guess which number and color the ball will land on when the wheel stops moving. Johnny explained that he had a complicated system that could help him predict which number the ball would land on . . . and then he proceeded to lose all his money.

While Johnny's calculations might not have worked out on that particular day, the Monte Carlo simulation is based on the game of roulette and is all about making predictions. The Monte Carlo simulation attempts a scenario—like rolling a ball on a roulette wheel—over and over and over again. Then it takes the average of the results to see if certain outcomes are more likely than others.

Today, computer programs use this method to analyze past data and predict future outcomes. It uses randomness to solve complex problems, such as predicting the weather (a task Klári helped with!), projecting stock prices, teaching a computer how to play chess, and even answering search engine questions.

The Monte Carlo method helped scientists like Klári figure out what happened inside a bomb. In Klári's simulation, her teams followed the numerous possible paths of individual neutrons inside the bomb. Using the Monte Carlo method, the scientists were able to predict the chances of the neutrons triggering a chain reaction. By using a computer program, scientists could safely create and run simulations to see whether the bomb would explode in real life—no safety goggles required.

And who ended up being able to communicate with ENIAC? Klári.

Thanks to Johnny's flowcharts, Klári was able to create a series of Monte Carlo simulations that turned physics problems into computer code.

After months of work, in April 1948, Klári executed her first program "run," as in "running a program." It was a success. This run marked the very first time modern-style code was executed on any electronic computer, and it proved that computers could be used to play out theoretical possibilities.

But there were obstacles along the way. On the second run, an error appeared in the code—a bug. Word got around, and people started doubting the reliability of any Monte Carlo computer simulation. The consequences were severe. Klári's coding error had put all the Monte Carlo work in jeopardy. The stress was getting to her, but she kept at it. In May 1949, Klári led a third run of Monte Carlo simulations; this one was a success.

Although Klári was making progress on the calculations, the world of computers was changing around her. By the summer of 1951, newer computers replaced the outdated ENIAC. By 1952, Los Alamos got its own computer, and the opportunity to work on it shifted from women back to men. Over time, Klári began to step back from leadership roles, though not before transferring her Monte Carlo codes to Los Alamos's new computer: MANIAC I (Mathematical Analyzer Numerical Integrator and Automatic Computer Model I). Then Klári's personal life changed dramatically. Johnny was diagnosed with cancer and died in 1957.

Klári left the world of coding behind, eventually remarrying and moving to California. She was offered jobs but never returned to work. In her spare time, she wrote her memoir, which never found a publisher. In 1963, after she had spent an evening socializing with friends, her body was found on the beach.

The why of Klári's death will always be a mystery, but the legacy of her contribution to computer science remains. She helped develop a language no one had used before that computers could understand. And along the way, she indeed led a memorable life. "It was sheer luck, and a strong tendency for rainbow-chasing, that made me a wanderer on two continents among this maze of people," she wrote.

# DR. ISABELLA AIONA ABBOTT

JUNE 20, 1919–OCTOBER 28, 2010

> Dr. Aiona Abbott's embodiment of science, culture, and passion makes her a science hero full of *kuleana* (responsibility for her community) and *mana wahine* (girl power).
>
> —Allison Lau, *The Ethogram,*
> UC Davis Animal Behavior Graduate Group

| **CLAIM TO FAME** | The First Lady of *Limu* |
|---|---|
| **WHY** | Considered the world's leading expert on central Pacific algae during her lifetime |

**FIRSTS**

* First Native Hawaiian to receive a doctorate in science
* First person of color and first woman to become a full professor of biology at Stanford University
* Leading expert in Pacific marine algae

For centuries, seaweed, which is called *limu* in Hawaiian, has held special significance for the Polynesian cultures of the Pacific region. This important marine life has a number of nutritional, medicinal, and environmental properties. Something Isabella Kauakea Yau Yung Aiona Abbott's family knew from the time she was a young girl.

Isabella was born in 1919 in Hana, Hawaii, to a Chinese immigrant father and a Native Hawaiian mother. Her Hawaiian name, Kauakea, means "white rain of Hana," though as she was growing up, she more often went by the Western name of Izzie. She was the second youngest and only girl among her eight siblings.

When Isabella was three years old, her parents moved the family from Hana, on the island of Maui, to Honolulu, which is on the island of Oahu, so that her brothers could attend high school in a larger city. It was from Hawaiian shores that Izzie received her first lessons in botany, the study of plants, which sparked a fascination that would last a lifetime.

Isabella's grandmother lived near the coast, and when her mother wanted to get the kids out of the house to release some energy, she would take them down to the water near their grand-mother's house. In Hawaiian culture, it was common practice for women to pass on the knowledge of seaweed from one generation to the next. So that's exactly what her mother did. Under her direction, Izzie and her younger brother would search for *limu* along the coast while their mother taught them the differences between the various species and swapped recipes with her friends.

"When you're looking for seaweeds, you're not exactly drowning, you know, or doing crazy things out in the water, so my mother and father both were happy to take us to the beach," Isabella said, recalling these joyful days foraging in the surf for interesting plant life. Her mother taught her all the names of the edible species. The ones she couldn't identify, she discarded as *ōpala,* the Indigenous word for garbage, remembered Isabella.

When not learning from the beach, Isabella attended the Kamehameha School for Girls. Isabella's keen interest in science further blossomed in the school's flower gardens. Unlike at the shore, the plants in the gardens were labeled with their Latin names. "That was the first time anybody told me that the scientific names meant something, just like the Hawaiian names meant something," Isabella said in an interview. In the gardens, Isabella helped grow and harvest beans that fed 150 girls who attended the school.

After graduating from high school in 1937, Isabella attended the University of Hawai'i. From the start, it was clear Isabella was in the right place. On her first day of classes, she met Donald Putnam Abbott, a young man from Chicago who was just as enthusiastic about the natural world as she was. The two sat side by side in botany class, and as the years went on, their fondness for each other only grew. In 1941, Isabella graduated with a degree in botany and Donald with a degree in zoology.

Isabella continued her education in botany and moved all the way to the University of Michigan to earn a master's degree in 1942. She and Donald got married in 1943. Donald had stayed at the University of Hawai'i for a time working as a professor, but in 1943 he enlisted in military service for World War II.

After the war, the couple was able to reunite at the University of California, Berkeley. In 1950, Donald received his PhD in zoology, and Isabella a PhD in botany, making Isabella the first Native Hawaiian—male or female—to receive a doctorate in science.

Despite their having received equivalent advanced degrees from the same university, it was clear the couple would not have equal career opportunities. Donald had multiple job offers from top universities: Columbia, Yale, and Stanford. He chose to teach at Stanford's acclaimed Hopkins Marine Station. Isabella, however, found herself struggling to find a job.

Stanford's esteemed Hopkins Marine Station sits on the Monterey Bay.

The pair moved from Berkeley to Pacific Grove, California, for Donald's job. At the time, Stanford had a strict policy against hiring two family members to work in the same department. There was no question that Isabella was a highly qualified candidate, but Stanford did not want it to appear that Isabella was hired because of Donald's position. So instead, Isabella focused her attention on raising their daughter, Annie, while she continued to study the algae flourishing along the California coast.

For Isabella, there was a very practical reason to learn about algae, one she had known since her mother first took her out on the beaches near her grandmother's house:

Algae is a nutritious source of food. "Hawaiians ate seaweed raw. It was cleaned and pounded and salt added as a preservative," said Isabella. "Many older Hawaiians, myself included, eat *limu* by itself because we like it. It has vitamins and minerals."

Not that Isabella loved all seaweeds equally. A few not in Isabella's good graces included *Gracilaria salicornia* and *Hypnea musciformis*. What did these seaweeds do that was so wrong? Well, it's not that they are bad plants. Rather, they are considered invasive species in the Pacific oceans. (Flora Patterson would likely have a thing or two to say on the subject.) Because the seaweeds do not normally grow in an area, when they arrive, they can often cause damage to the ecosystem. *Gracilaria salicornia,* for example, smothered Hawaii's reefs, and *Hypnea musciformis,* also known as hookweed, damaged native algae. "That's not a hospitable way for a visitor to behave," Isabella pointed out.

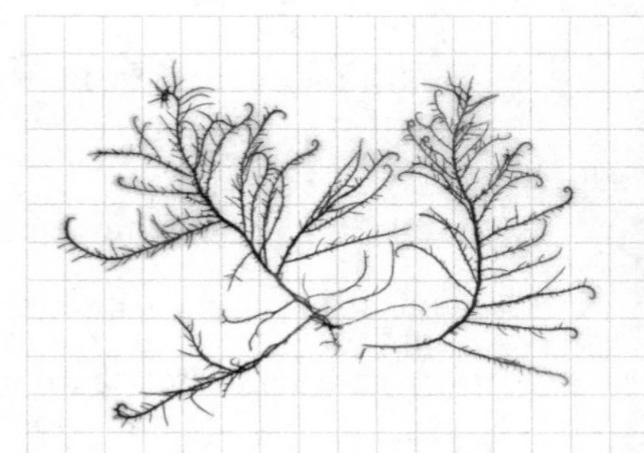

## Seaweed's Superpower

The next time you tuck into a sushi roll, consider that you may be biting into one of the planet's defenses against global warming: seaweed. Although it is called a weed, seaweed is not a plant, but a type of algae that grows in marine habitats. Scientists believe that seaweed—in addition to containing essential nutrients like vitamins A, C, and E, and minerals like iodine, calcium, and iron—is also a superhero in the fight against climate change.

Upwards of 9,000 different types of seaweed have been identified, many by Dr. Isabella Abbott, and the number increases all the time.

There are three main groups of algae: red algae (Rhodophyta), brown algae (Phaeophyceae), and green algae (Chlorophyta). Unlike plants, algae lack roots, stems, and leaves. However, like plants, algae can use photosynthesis, a process that converts the sun's energy into sugars. Instead of using leaves to collect sunlight, algae have multicellular structures called thalli that absorb the powerful rays.

During photosynthesis, algae also absorb and store vast amounts of carbon dioxide, or $CO_2$, from the atmosphere. Brown algae are particularly effective at storing the gas. When there is too much $CO_2$ in the environment, it can negatively impact marine ecosystems and make the oceans acidic. Seaweed helps maintain the acidity, or pH balance, in oceans, allowing plant and animal life to thrive. In addition, seaweeds need nutrients like nitrogen and phosphorus to grow, so they often can make good use of agricultural runoff and pollution that may harm other plants.

It also helps that seaweed forests can absorb up to twenty times more $CO_2$ per acre than trees in terrestrial forests. This is not only because algae grow faster, but also, thanks to the depth of the ocean, because dead plant material sinks and stays buried longer, which releases $CO_2$ at a slower rate.

And it's not just the oceans that benefit from seaweed's pollution-fighting powers. Seaweed is an effective reducer of methane gas, which can be created during cattle farming and fossil fuel use, along with other activities that contribute to global warming. Compounds in certain species of seaweed actually reduce methane production in cows' digestive tracts. So by simply adding seaweed to cows' diets, greenhouse gas emissions could be reduced by a noticeable level.

Lastly, seaweed can also be used to produce biofuel, which is fuel made from plant material or animal waste. Unlike fossil fuels, which include petroleum, coal, and natural gas, biofuels can be replenished more easily, making them a renewable resource.

As weeds go, this is one that should be welcome in Earth's ocean gardens.

Isabella's thoughtful work continued to be noticed. Starting in 1956, she was able to skirt Stanford's no-spouses rule by getting a job as a lecturer at Hopkins Marine Station. A lecturer is not the same as a professor; Isabella would not yet receive the same academic recognition as Donald, but this position was a way for her to teach. She also continued to study different types of algae along California's Monterey Peninsula, publishing papers on her discoveries. Isabella coauthored several books during this time, and in 1969, the Botanical Society of America awarded her the Darbaker Prize "for her taxonomic and morphological studies of marine red algae of the Pacific northwest coast of North America." In 1972, Stanford University had no choice but to recognize Isabella's tremendous contributions. They promoted her to full professor of biology, marking the first time a person of color or a woman was hired at the university to work in biological sciences at such a lofty level. The position previously had been held exclusively by white, male faculty members.

Isabella and her husband worked at Stanford for ten more years. During their time at the university, they became recognized as two of the top voices in their field, passionate teachers, and strong mentors for their students. In 1982, Isabella and Donald decided it was time to retire from Stanford and move back to Hawaii. But Isabella was far from finished.

Soon after the move, Isabella became a professor of botany at the University of Hawai'i at Mānoa, where she established the university's first undergraduate major in ethnobotany, or the study of how humans developed plant lore, often with regard to Indigenous cultures. This accomplishment was significant because Isabella was not only adding to the world's knowledge of marine life, but she was also highlighting and preserving the Indigenous Hawaiian culture that she loved so deeply. Much of her work was based on oral histories. She interviewed elders in the community to uncover and record historical uses for marine algae, thus preserving the vital link between the Hawaiian flora and culture, from the past to the present.

## The Ins and Outs of Ethnobotany

The years Isabella Abbott spent learning about *limu* from her mother impressed upon her the vital link between food and family. No

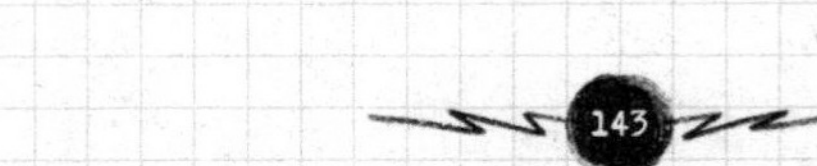

wonder she became a champion of ethnobotany. This branch of science combines anthropology, the study of humans past and present; botany, the study of plants; and ecology, the study of the relationships between living organisms and their environment.

Humans have always depended on plants to survive. Over the centuries, cultures have experimented with vegetation and passed down food, clothing, shelter, religious, and medicinal uses that they discovered from plants so that future generations could benefit. In fact, many modern medicines originated from traditional plant-based remedies.

Ethnobotany explores how specific cultures interact with the native plants in their region. Ethnobotanists work with Indigenous communities to document their traditional practices, knowledge, and beliefs about plants and work to preserve their customs so that both their culture and their ecosystems are sustained.

Establishing this history is also crucial for coming up with new solutions. As communities struggle with food insecurity and the destruction of their natural environments, Indigenous populations are often already using plant species to combat these challenges. Together, Indigenous communities and ethnobotanists have found ways to incorporate these established practices to create new goods, including plant-based meat alternatives and dairy-free milk and ice cream; clothing made from cotton, hemp, flax, and bamboo; and even products like makeup, soap, and chewing gum.

Isabella also created a hands-on ethnobotany lab so that students could directly interact with Hawaii's native plant and marine life and put the traditional practices into use. It's said that students would stand in crowded lecture halls and listen in the aisles so that they could have the opportunity to learn from Dr. Abbott.

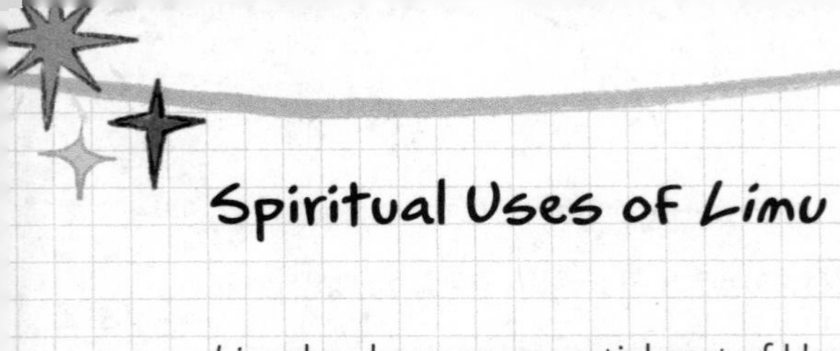

## Spiritual Uses of *Limu*

*Limu* has been an essential part of Hawaiian culture for centuries. The species *limu kala* is an important food source for both humans and turtles. But its uses don't stop there. The word *kala* means "to forgive," and this seaweed is used in spiritual practices and important events in people's lives, including the preparation before embarking on a dangerous journey and the Hawaiian reconciliation process known as *ho'oponopono.* According to the Hawaiian mythic poem called the Kumulipo, all Hawaiian life sprang from the coral polyps in the sea, and *limu* in particular is considered sacred and believed to possess healing properties. *Limu* is associated with Hawaiian gods and goddesses and represents abundance, protection, and fertility. Ancient Hawaiians believed that *limu* connected them not only to their planet but also to their ancestors.

And as seaweed gained popularity in cuisines around the world, Isabella recognized the importance of maintaining the traditional way of harvesting *limu* and sharing it with others. Gathering *limu* according to the Indigenous practices required careful handpicking so that the algae would grow back. Isabella taught others this form of sustainable harvesting so that the coastal ecosystem could be preserved for generations to come.

She also sought practical applications for cultivating and harvesting seaweed so that it remained a useful part of everyday life. Her book *Lā'au Hawai'i, Traditional Hawaiian Uses of Plants* was the first comprehensive description of the ways Hawaiians use plants for food, clothing, shelter, and tools, and in religion and recreation, like crafts.

TRY IT!

## Bake Your Own Seaweed Cake

Isabella's recipe for seaweed cake, which she often baked for potluck dinners, is from her book *Limu: An Ethnobotanical Study of Some Hawaiian Seaweeds.*

*This recipe was originally designed to use Nereocystis kelp that is common in central California and northward. In Hawaii, either Eucheuma species from Kāne'ohe Bay, or ogo may be used. (Wakame, which means "young girl" in Japanese, can be substituted in this recipe.)*

* Cream well 1½ cup salad oil and 2 cups sugar; add 3 eggs, one at a time, beating well after each addition.
* Add 2 cups grated carrots, 2 cups grated Eucheuma or 2 cups coarsely chopped ogo (or substitute with

unseasoned seaweed from the grocery store), and 1 cup crushed, drained pineapple or 1 cup fresh grated coconut.

* Sift together $2\frac{1}{2}$ cups sifted flour, 1 teaspoon baking soda, 1 teaspoon salt, 1 teaspoon cinnamon. Mix all together.
* Add 1 cup walnuts if desired.
* Bake in an oblong pan or loaf pan at 350 degrees 45 to 50 minutes.
* Serve plain or with buttercream frosting. This moist cake keeps very well.

Isabella remained an active scientist until she died in 2010 at the age of ninety-one at her home in Honolulu. During her lifetime, she wrote 8 books and more than 150 articles about algae and seaweeds, and is credited with discovering over 200 different algae species. Several species are named after her, including Abbottella, which means "little Abbott." Today, several of her specimens are part of the Smithsonian's National Museum of Natural History's botany collections.

Isabella also won numerous awards, including the Charles Reed Bishop Medal; the National Academy of Sciences Gilbert Morgan Smith Medal, for her research on algae; and a lifetime achievement award from the Hawaii Department of Land and Natural Resources, for her studies of coral reefs. To honor her contribution, the University of Hawaiʻi established a scholarship to support graduate research in two areas closest to Isabella's heart: marine botany and Hawaiian ethnobotany. And in 2022, Stanford's Hopkins Marine Station dedicated a lecture hall in Isabella's honor as a testament to her pioneering work and lasting legacy.

Of all of her accomplishments, though, Isabella said that being named one of the Living Treasures of Hawaiʻi by the Buddhist organization Honpa Hongwanji Mission of Hawaiʻi "moved me the most, because it's for your contribution to your community." Indeed, from Isabella's first lessons on the beach with her mom, she spent her life studying, documenting, and sharing her knowledge of seaweed while also protecting the ecosystem. Her work embodies the spirit of *aloha*—the Hawaiian word for love, peace, and compassion.

# YVONNE Y. CLARK

APRIL 13, 1929–JANUARY 27, 2019

> "Now you might call me a pioneer, but I don't see myself as one. I am tired of being the first person to do something. Where are the rest of these people?"
>
> —Yvonne Y. Clark

| CLAIM TO FAME | The First Lady of Engineering |
|---|---|
| WHY | For her groundbreaking achievements as a Black female mechanical engineer |

**FIRSTS**

* First woman to earn a bachelor's degree in mechanical engineering from Howard University
* First woman to earn a master's degree in engineering management from Vanderbilt University
* First woman engineer hired as an instructor in the engineering department at Tennessee State University
* First Black member of the Society of Women Engineers

From the get-go, Yvonne Clark, known as Y.Y., was a troubleshooter. When she was nine, her family's toaster broke. Without telling anyone, Y.Y. snuck the appliance out of the garbage, took it up to her room, and set to work. With pliers from her Erector Set and two paper clips, Y.Y. dismantled the appliance and experimented with adjusting the release mechanism, discovering that the spring was bent, which was causing the bread to get stuck and burn. She persisted in her fiddling until she fixed it. Once everyone in her house had gone to bed, she snuck the toaster back downstairs to the kitchen. "I can't express the sense of accomplishment I felt having successfully solved a real-world problem," she said. "I felt like I did on Christmas Eve in excited anticipation of what the next day held."

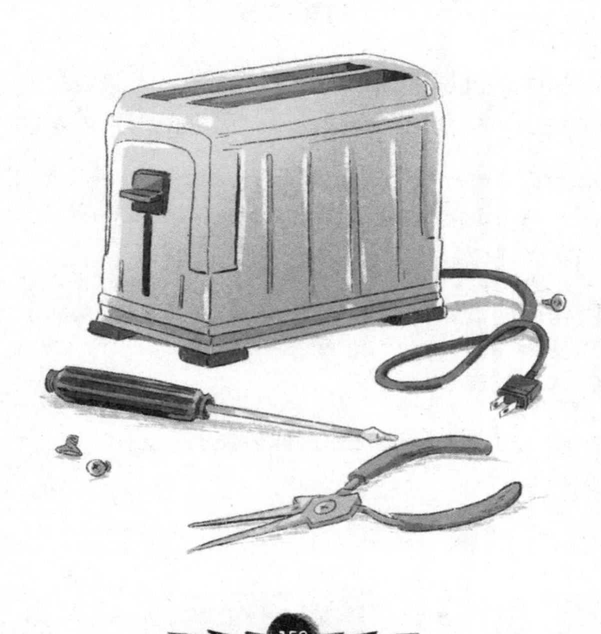

The following morning, Y.Y.'s mom woke her for breakfast. When Y.Y. came downstairs, she soon saw that awaiting her was perfectly toasted bread. The toaster worked so well that their housekeeper thought it was a brand-new one. Her father knew otherwise and gave his daughter a stern talk on fire safety!

Still, Y.Y.'s parents were so proud that they used the money they would have spent on a new toaster on a special gift instead—a silver charm bracelet adorned with a single charm of a toaster. "You can use this bracelet to put charms for all of your accomplishments, but we wanted to give you the first one," her father said.

Y.Y.'s special knack for figuring out how things worked, and her dogged, never-give-up approach, would lead to many accomplishments in her life, including significant achievements in academics and breakthroughs in engineering. But before she could do either, Y.Y. would have to knock down the barriers she faced because she was both Black and female.

## Words Worth

As smart as she was, not everything came easy to Y.Y. She struggled with speech and stuttered when she was a little girl. Stuttering means someone repeats sounds, syllables, or words while trying

to speak or has trouble expressing their thoughts. Because of her stuttering, Y.Y. was treated unkindly by her classmates, and eventually she became so self-conscious that she almost stopped speaking altogether.

But Y.Y. was not alone. Stuttering, also known as stammering, afflicts more than 80 million people worldwide, which is 1 percent of the population. Stuttering is four times more common in men than women. Many famous people have stuttered, including President Joe Biden, golfer Tiger Woods, basketball player Shaquille O'Neal, actresses Emily Blunt and Nicole Kidman, and Lewis Carroll, author of *Alice's Adventures in Wonderland.*

Different things can cause stuttering, including genetics—which means that it is passed down from parents—a developmental delay during childhood, or even family dynamics, where there are high expectations and pressure to keep up. About 5 percent of children go through a short period of stuttering. The majority of people outgrow the condition, especially when it is treated early with speech therapy.

The best thing you can do when someone is stuttering is to be patient and let them finish what they are saying without jumping in. Try to focus on *what* the person is saying, rather than *how* they are saying it. The more relaxed you can be as a listener, the more at ease the person will be as a speaker.

As anyone who has suffered setbacks or learning differences

knows, there are benefits to facing challenges in life. For Y.Y., in her own words, it meant developing "rhino skin." The experience of being different toughened her up. This would come in handy when she kicked down all the barriers in her way because she was Black and a woman. She also learned how to slow down and be clear so that others could understand her, which helped her solve and explain complex problems.

Y.Y. had some things in her favor. Her family was well educated and prosperous. Y.Y. and her younger brother grew up in the 1930s in an upper-middle-class Black neighborhood in Louisville, Kentucky. Their father was a doctor. Their mother was a librarian and newspaper columnist. Her parents were supportive of her unusual talent in engineering, which is the method of using math and science to solve problems, even though it was a field dominated by men.

The world that Y.Y. grew up in, however, was not receptive to her talents. Like the rest of the South, Louisville was segregated, which means that Black and white people were kept apart, with Black people being treated unfairly in every area of their lives—from how they were educated to how they were treated in society. (You may remember that medical trailblazer Dr. Sarah Loguen Fraser fought against similar unjust laws in her lifetime.)

In Kentucky, Blacks were forbidden by law from attending the same schools as white children. In public, Black citizens had to use separate bathrooms and separate water fountains and sit in different sections of restaurants and on public transportation.

Y.Y. attended an all-Black high school in the early 1940s, when the United States entered World War II. The military sent many male pilots overseas, so the US Army Air Forces began to recruit women. The female pilots were known as the Women Airforce Service Pilots, or WASPs, and Y.Y. wanted to join. She graduated from high school two years early with her sights set on becoming a pilot, but the WASPs did not accept Black women.

Y.Y. knew that the discrimination wasn't just morally wrong, it was also illogical. Why should race or gender prevent a qualified person from doing a job, especially one that needed doing?

Still, Y.Y. was determined to use her mechanical skills in the field of aviation. In 1947, she was accepted by the University of Louisville, an all-white college at the time. However, once the school learned that Y.Y. was Black, they said it was against the law for them to admit her, which was unfortunately true.

But this law was so unfair that Y.Y.'s parents decided to step in and take legal action against the school. In the end, the University of Louisville agreed to pay for Y.Y. to go to Howard, a historically Black university in Washington, DC, that offered mechanical engineering classes, if her family agreed not to sue the university. So it was decided. Y.Y. went to Howard, but even there, more often than not, she was the only woman in her classes. But she refused to let that get in her way.

## Hooray for Historically Black Colleges and Universities

When Y.Y. was going to college in the 1940s, Black students were not allowed to enroll in many white colleges. In the South, Blacks were

legally barred from white colleges. Even in the North, colleges and universities limited the number of Black students who could enroll. Luckily for Y.Y., there was another option to further her education. Historically Black colleges and universities, also known as HBCUs, were established expressly to offer the basic human right of higher education to Blacks and other minorities.

The first HBCUs were founded before the American Civil War (1861 to 1865) to teach the children of formerly enslaved people. Pennsylvania and Ohio were home to the first two HBCUs, and their purpose was to teach Black students to become tradespeople and teachers. After the Civil War, during the period known as Reconstruction, the federal agency called the Freedmen's Bureau helped found a number of HBCUs, including Morehouse College and Spelman College, both in Georgia. Booker T. Washington, a former enslaved person who became a famous author and champion of education, founded the Tuskegee Institute in Alabama. Tuskegee focused on careers in agriculture, which is the science of cultivating soil, growing crops, and raising livestock.

In 1964, the Civil Rights Act prohibited discrimination on the basis of race, color, religion, sex, or national origin. This made it illegal for colleges to reject students because they were Black.

Today, there are more than one hundred HBCUs, mostly in the South. Just as with other colleges and universities, there are both private and public HBCUs.

Howard University in 1910.

Today, nearly 230,000 students are enrolled in HBCUs. While they were first established for Black students, now one in four students attending an HBCU is not Black. Still, these schools remain special in that they provide a welcoming environment that honors the history, perspective, and experience of Black Americans. Famous graduates include Dr. Martin Luther King Jr., Oprah Winfrey, and Vice President Kamala Harris.

Y.Y. became the first woman to earn a degree in mechanical engineering at Howard. But instead of being honored for her achievement, she was informed that because she was a woman, she wouldn't be allowed to participate in the graduation ceremony.

After college, she was rejected for jobs because of bias against both Blacks and women. But Y.Y.'s persistence paid off. At her first job, helping to make weapons for the US Army, Y.Y. fixed a cannon that jammed up in cold weather. Several teams of men had already tried, and failed, to determine the cause.

Y.Y. took the diagrams for the cannon and blew them up to four times their original size so she could see everything clearly. This was before computers, and not an easy task. But it worked: Y.Y. zeroed in on the detail that showed where two pieces of metal in the mechanism were overlapping, causing the jam. Identifying what was causing the problem was the first step in solving it.

Y.Y.'s problem-solving approach was stunning in its simplicity. She understood how important it was to take a step back and see things clearly before forging ahead. Thus began her reputation as a professional troubleshooter who could fix problems that others couldn't.

Y.Y. went on to other jobs in engineering. When she was twenty-six, she married William F. Clark Jr., a biochemistry teacher. Y.Y. moved from New Jersey to Nashville, Tennessee, in the segregated South to be with her husband. With few options in engineering for Black women, Y.Y. took a job teaching mechanical engineering at Tennessee State University in 1956, becoming the first female faculty member in the department. That year, she also gave birth to her son, Milton.

While on her summer breaks, Y.Y. worked for a company that had a contract with NASA. She spent her time on two big projects: the F-1 engines of the Saturn V rocket and the Apollo Lunar Sample Return Container, which was used to collect rocks and other samples from the moon. The race was on between the US and the Soviet Union to put the first man on the moon. The US was hoping to make this happen with the Saturn V rocket, which was the largest rocket ever built. But the rocket had a problem: Its engine got too hot.

Y.Y. used the same logic she had employed earlier in her career with the cannon: If a problem seemed impossible to solve, then someone wasn't looking at it right. First Y.Y. analyzed the mathematics and the design of the rocket. When she couldn't find the flaw, she decided to do some firsthand reporting. She called the team in Florida that was running the tests and asked them very specific questions. She got some very specific answers: It turned out the technician in the field had forgotten to put on the covers and tighten the wires. Bingo! Y.Y. discovered that the sensor, which registered the heat, had been installed incorrectly, a crucial element that all the prior team members had missed. That answered the question of why the sensor was registering the high temperatures.

On July 16, 1969, the *Apollo* 11 Saturn V rocket launched with astronauts Neil Armstrong, Michael Collins, and Buzz Aldrin on board en route to the moon.

To Y.Y., the right answer had to start with the right question. She didn't solve the *problem;* she solved the *question.* In this case it was: What was causing the sensor to go off? Y.Y. had a gift for reducing seemingly complex problems to their essence.

Then, in 1968, Y.Y.'s daughter, Carol, was born. Commuting between her home in Tennessee and a job in Alabama would be too much with two children. Y.Y. gave up her work with NASA, instead becoming the director of a program at TSU that sponsored students for internships at NASA. Y.Y. continued her education as well, becoming the first woman of any race to earn a master's degree in engineering management from Vanderbilt University. By that point, the restrictions around segregation were easing up.

## A Rock-Solid Plan

From 1969 to 1972, *Apollo* crews returned 2,200 samples from six landing sites on the moon. The boxes that were built to carry the precious cargo harvested over the course of six missions were designed in part by Y.Y.

Known as rock boxes, these special containers looked like aluminum carry-on suitcases. Beneath the basic exterior, the boxes had a high-tech engineering design to safely bring the moon samples back to Earth.

First, the rock boxes had to accommodate dramatic temperature changes. Because the moon has no atmosphere, temperatures range from as low as -410°F to a scorching 250°F. The design team had to figure out how to keep temperatures stable in the rock box. To deal with the cold, they decided to use a highly polished aluminum surface. To deal with the heat, their solution was to make these rock boxes reflective.

The second challenge was ensuring that there would be a vacuum inside the rock box. A vacuum is a space devoid of any matter—there's nothing there. To prevent contamination, the box had to be vacuum-sealed before it left Earth. Even air couldn't get inside, or the box might become pressurized (like a hissing soda can) when the astronauts opened it up in space. For this, the team designed gaskets with O-ring seals like the silicone ones you might find on a reusable water bottle.

The astronauts then had to be able to vacuum-seal the box on the moon. This was tricky because the bulky space suits and gloves limited the astronauts' dexterity. The design team came up with a solution: to seal the box by simply closing it. For this, they created what's called a knife's-edge seal. When the astronauts closed the box, the rigid blade would slice into the softer metal, like a knife cutting through a stick of butter, ensuring that the box would have an airtight seal.

Finally, to prevent the rocks from getting crushed by vibration as the spacecraft hurtled back to Earth, the box was lined with aluminum mesh packing material. Thanks to Y.Y. and the team of engineers, nearly 850 pounds of rocks, dust, pebbles, core samples, and soil made the trips from the moon to Earth safe and sound.

Y.Y. spent fifty-five years at Tennessee State University. During that time, she helped bring many more women into the mechanical engineering—until women represented 25 percent of the students in that program. Student by student, engineer by engineer, Y.Y. transformed her field. And she didn't do that just by being the first—she did it by educating and inspiring and nurturing the next generation.

By the end of her life, Y.Y. could have had a charm bracelet brimming with symbols of her accomplishments. The secret to her success can be summed up in her no-nonsense advice: "Don't take no for an answer if you feel you're qualified," she said. "Keep your respect. Respect others. And just don't give up."

As co–executive producer of *Lost Women of Science,* I've sought to champion the stories of female scientists. I've marveled at their intellectual curiosity, celebrated their triumphs, mourned their losses, and deeply admired their strength—all while recognizing how strong they had to be simply in order to be heard. Little did I know that one of these overlooked scientific trailblazers was someone in my very own family. The last biography in *Lost Women of Science* belongs to my grandmother: Dr. Leona Zacharias.

—Katie Hafner

# DR. LEONA ZACHARIAS

FEBRUARY 1907–JANUARY 1990

“I had already been on a mission to shed light on female scientists who never got the credit they deserved when they were alive. Much to my surprise, it turned out that my grandmother was one of them.

—Katie Hafner, co-executive producer of *Lost Women of Science*”

| **CLAIM TO FAME** | The Visionary Biologist |
| --- | --- |
| **WHY** | For providing essential research in discovering why premature newborns were going blind |

**FIRSTS**

* Only female research fellow, a prestigious faculty position, at Columbia University in 1936
* Only woman to earn a PhD in anatomy in her graduating class

In the 1940s, in rich countries, also known as industrialized countries like the United States, premature newborns started going blind shortly after being born with perfectly healthy eyes. The mystery baffled doctors. Yes, the babies were born early and weighed less than four pounds, but that alone shouldn't cause sudden loss of sight. A team of scientists came together to investigate, including Dr. Leona Zacharias—who also happens to be my maternal grandmother.

Growing up, I never knew how important my grandmother's scientific work was because it was always my grandfather's career that took center stage. My grandfather Jerrold Zacharias was a prominent nuclear physicist. In the 1950s, he was a science advisor to President Eisenhower. He also developed the first atomic clock, which revolutionized timekeeping in the scientific field. My grandfather's work was the legend our family discussed, our main source of collective pride. My grandmother's career? It never came up.

To be honest, my grandmother was not the warm and fuzzy type. I spent my childhood in fear of her. She was cold and quick to judge. To this day, I remember the hot sting of her words and how every morning she would cut toast into precise squares. Whenever my sister and I visited, we would fight over the privilege of tossing the morsels into the air for her beloved poodles to catch. Of course, we were seeking our grandmother's approval.

When I was about seven, my family was visiting my grandparents, and one morning I woke up with an eye infection called conjunctivitis, also known as pink eye because it makes your eyes red and swollen. Both my eyes were nearly crusted shut because of the goop that was leaking out of them from the infection. To my surprise, it was my grandmother who nursed me back to health. She tenderly applied ointment twice a day, held my hand, and talked my fears away. While that gentleness did not last, it always stuck with me. It was only many years later, as an adult, that I discovered the reason for her compassion purely by coincidence.

Leona Ruth Hurwitz was born in New York City in 1907, to a middle-class Jewish family with Eastern European roots. Her father, my great-grandfather, was a high school math teacher. Leona's parents encouraged her to pursue college and a career, which was unusual advice for a young woman at the time. Mostly, women were expected to get married and have children.

In 1925, when my grandmother was a sophomore at Barnard College, she met Jerrold Zacharias, a senior at nearby Columbia University, who was studying physics. He was from Florida, funny, and wealthy. They made a handsome pair.

Two weeks after Leona graduated from Barnard with a degree in biology, she married Jerrold. She was twenty. Leona continued her education, earning a master's degree in zoology from Columbia, and shortly after graduating, the couple welcomed their first daughter, Susan, my mother. While raising a young child, Leona completed a PhD in anatomy, also from Columbia. This was quite an accomplishment.

My grandmother was definitely ahead of her time. In 1936, she was the only female research fellow, which is a prestigious faculty position, at Columbia. A couple of years later, of the fifty-four science PhDs awarded, only eight went to women.

And she was the only woman at Columbia that year to receive a PhD in anatomy. Still more impressive, a prominent science journal published her PhD dissertation, which is a long research paper that sheds new light on an important topic.

My grandmother was a very hard worker. At one point early in her career, she held three jobs simultaneously, all while raising a young child and being a supportive faculty wife to my grandfather, who was a rising star in the Columbia physics department. She kept working through the birth of a second daughter (my aunt Johanna). In 1945, when my grandfather got a job at the Massachusetts Institute of Technology, my grandmother and the two girls followed him to Boston. At Harvard Medical School, my grandmother got a job as a lecturer in ophthalmic research, which is the study of eyes. It was a position that would come to define her career.

As it turned out, she walked straight into one of the most confounding medical mysteries at the time. Babies were going blind, and no one knew why. The blindness affected infants who had been born prematurely—meaning they were delivered before their mothers completed a full nine months of pregnancy and they were very small. Prior to the 1940s, many premature babies died because their bodies had not developed fully enough to live outside their mothers' wombs. New technologies, though, such as incubators, were now helping keep premature babies alive by providing an environment where the baby could have a controlled amount of oxygen, humidity, and light while the vital organs developed.

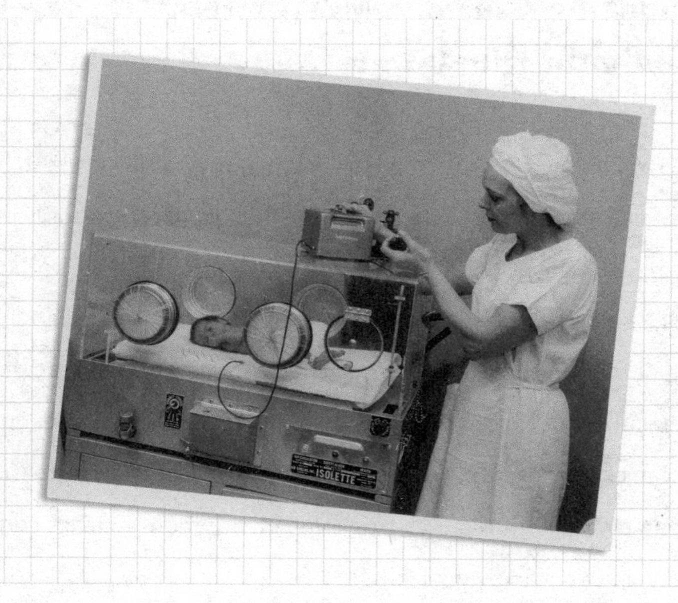

However, as these new technologies were integrated into hospital practice, doctors noticed that a disturbing number of these premature babies were going blind. Doctors first called this strange occurrence retrolental fibroplasia. (Later it became known as retinopathy of prematurity, or ROP.) The unusual eye disease was caused by abnormal tissue forming behind the lens of the eye, resulting in unclear vision and in some cases total blindness. By the late 1940s, the mysterious syndrome had exploded into a full-blown epidemic. And it was happening all over the world in developed countries where hospitals could afford incubators.

## Debunking Common Myths About Childhood Blindness

There are many myths floating around about what it means to be blind. For instance, a lot of people believe that blindness means living in total darkness. But that is not the case. Most blind individuals can see some light and shadows.

Another myth is that blind people automatically develop other superacute senses, like hearing, to compensate. Although people who can't see may develop enhanced sensory perception, it doesn't happen by magic. It requires training and intention, which is completely up to the individual.

And some even believe that people with visual impairment don't dream. But they most certainly do! It's just not always in pictures. For those who have been blind since birth, their dreams will draw from things they have heard, touched, smelled, tasted, and felt. According to a study conducted by the National Federation of the Blind, the content of dreams does not depend on visual experiences, but rather on memories, personal encounters, and even imagination.

Nearly 3 percent of children younger than eighteen are blind or visually impaired. An estimated 1.4 million children under the age of fifteen are affected worldwide, and a disproportionate number of those impacted live in more economically disadvantaged countries.

Vision loss can occur for a multitude of reasons: accidental, environmental, or genetic—which means passed down from parents. The leading cause of childhood blindness is cataracts, a clouding of the normally clear lens of the eye, which is caused by a vitamin A deficiency. For children afflicted, it's like looking through a blurry or foggy window. The second leading cause is retinopathy of prematurity (ROP), which affects premature infants. With ROP, the blood vessels in the retina (the light-sensitive layer in the back of the eye) don't develop normally. The third leading cause is glaucoma, a progressive eye disease that affects the optic nerve, which transmits information to the brain. With all of these diseases, early detection and surgery can often reduce the severity.

To better understand visual impairment, it's helpful to know how we see. Vision uses a combination of the eyes and brain. Your eyes have different parts, including the retina, lens, iris, and cornea. Together, these parts detect light, shapes, and colors, and send the information to the brain through special nerves. Instantly, the brain processes the information so that you recognize what you're looking at. When one or both eyes are damaged, this process breaks down, leaving a person a skewed vision of what they are seeing, what we call visual impairment or blindness.

Being blind doesn't mean losing your independence. People who are blind can often live on their own with the aid of different strategies and adaptive technologies such as a cane or a guide dog to help with mobility. Screen readers, braille displays, and accessible smartphone apps can also help with communication and accessing information. Home modifications may include installing talking appliances and voice-controlled systems to make operating common devices easier. It also always helps to have a community to listen and provide resources and support—something we all need from time to time. Thanks to these and other accommodations, many blind children have grown up to achieve amazing feats in various fields, including art, music, sports, and more.

In 1948, researcher Theodore L. Terry published a scientific paper outlining the history and theorizing about probable causes of the blindness. Theodore was a prominent ophthalmologist at Mass Eye and Ear, a Harvard teaching hospital dedicated to eye, ear, nose, throat, head, and neck care and surgery. His paper—of which he is credited as the sole author—tests a number of hypotheses about the causes of the disease: birth order, sex of the baby, deficiencies in iron or vitamin A, and oxygen, which was routinely given to premature infants with underdeveloped lungs.

By the early 1950s, researchers in at least a dozen different locations across the world were sharing their findings. The results pointed conclusively to one culprit: oxygen.

Incubators were growing more sophisticated, allowing for higher levels of oxygen concentration. Oxygen was beneficial for premature babies because their lungs were too immature to breathe on their own. Unfortunately, too much oxygen was causing irreversible problems. It turned out that if a baby's eyes receive excess oxygen while the baby is in the incubator, the blood vessels in the back of their eyes stop growing normally. Once the baby leaves the incubator, their eyes try to make new blood vessels, but the action strains the retinas and can cause them to detach from the optic nerve, or the bundle of nerves that connects the eye to the brain. This detachment was causing the babies to lose their eyesight. How can we be sure? Because as soon as new standards were put in place to reduce oxygen levels in incubators, the incidence of blindness decreased sharply.

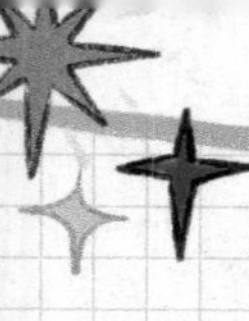

# The Musical Wonders of Stevie Wonder

Superstar Stevie Wonder is one of the most influential and creative musical artists of our time. His remarkable talents as a singer, songwriter, and multi-instrumentalist have led to numerous hit songs and albums and a career spanning decades. His soulful voice, innovative music, and unique talent for blending rock, jazz, and pop have earned him twenty-five Grammy Awards. And he has made all these significant contributions to the music industry as an artist who is blind.

Born six weeks early in 1950 in Saginaw, Michigan, Wonder was put in an incubator, as was then the custom to help premature babies breathe. Wonder developed retrolental fibroplasia, the very disease Dr. Leona Zacharias and others were studying at the time. In Wonder's case, the disease led to blindness.

But that didn't stop him from sharing his astounding gift of music with the world. From an early age, Wonder's talents were apparent. Before he was ten, he had taught himself how to play the piano, harmonica, and drums. When his mother moved the family to Detroit to get away from his abusive father, Wonder met a musician named Ronnie White who was in the band the Miracles. White was so impressed with Wonder's talent that he introduced him to Motown Records founder Berry Gordy, Jr., who was the most influential producer in the music business at that time.

From there, Wonder had a rocket ride to stardom. His hit songs "Superstition," "Sir Duke," and "Isn't She Lovely" ruled the airwaves in the 1970s and 1980s.

In addition to his musical prowess, Wonder has also been a lifelong champion of social justice. He pushed for Martin Luther King Jr.'s birthday to become a national holiday and wrote and recorded the joyful anthem "Happy Birthday" as a tribute to honor the civil rights leader. When his hit "I Just Called to Say I Love You" won the 1985 Oscar for Best Original Song, Wonder took the opportunity to publicly dedicate the award to antiapartheid activist Nelson Mandela.

Wonder has been celebrated with many awards, including induction into the Rock & Roll Hall of Fame, a Grammy Lifetime Achievement Award, and the Presidential Medal of Freedom.

When questioned during an interview about how his lack of sight affected his music, Wonder had this to say: "I'm able to use my imagination to go places, to write words about things I've heard people talk about. In music and in being blind, I'm able to associate what people say with what's inside me."

I didn't learn of my grandmother's role in this great medical mystery until 2022. Our teams had launched the *Lost Women of Science* podcast earlier the previous year, and I had been steeped in research about other women whose accomplishments were dimmed or entirely overlooked by the scientific community. It struck me that my grandmother had also been a scientist. Could it be possible that she was one of these lost women, too?

In the fall of 2022, on a whim, I did an online search of her name and was amazed to find her listed in the archives at Harvard and MIT. Naturally, I thought for her name to appear in the special collections at such institutions must mean that her research contained something worth investigating.

A few weeks later, I was off to Boston to see the collections for myself. Over the course of two days, I combed through the boxes, and I ended up locating Dr. Theodore L. Terry's 1948 paper on the infant blindness epidemic. His was the only name on the front cover, but I knew my grandmother's slightly backward-slanting handwriting well, and I spotted it on a number of pages. She was charting, graphing, and tracking every variable that might show a pattern to explain the blindness. In one experiment, she even tested the effects of vitamin E deficiency in baby chicks.

Learning about the hunt for answers that my grandmother worked on was inspiring, but I couldn't ignore something that bugged me while I went through the files. Theodore Terry died two years before his article was published. Even more curious was the fact that his original paper included, handwritten in red ink, my grandmother's name, followed by "with best wishes to the real author," penned by someone named Everett. Did that mean my grandmother actually wrote Terry's paper after he died? And who was Everett?

I had to know more. As I continued my search in the very building where my grandmother had her office at Mass Eye and Ear in

Boston, I located a 1949 report written by my grandmother and an Everett Kinsey. After Terry's death in 1946, I suspect that Everett probably took over the lab and became my grandmother's new boss. And it appears that Everett and Leona became a real team.

After finding out how much work my grandmother had put into trying to find the cause of blindness in so many babies, I got to thinking that she really didn't get the credit she deserved. Dismayed, I took one more look at the report, and that's when I saw it. There, at the bottom of the front page, in a small typeface, is a tiny asterisk, and then this: "Edited after Dr. Terry's death by a committee." The first name listed on the committee: Dr. Leona Zacharias.

There was my grandmother's credit! In the end, she was acknowledged, but that asterisk is a symbol for the way each of the scientists in this book and many women's contributions in my grandmother's era were acknowledged, which is to say hardly at all.

Here's the thing: History tends to simplify scientific breakthroughs by giving the credit for a discovery to one person, usually a white man. But discoveries are rarely made by just one person working alone; they usually rely on a multitude of foundational scientists and their years of questioning. And it is clear that my grandmother's many years of probing constituted one of those parts.

But the tremendous discoveries, efforts, and proof of pioneering ideas from women are there—if you look hard enough. The people with asterisks are not the ones who get awards, fame, or promotions. They often aren't asked to give speeches, or become deans or department chairs or even full professors. Even the fact that I missed that tiny asterisk the first time I read the paper shows how easily women's contributions can be overlooked.

That only makes it more important that we seek them out.

Dr. Leona Zacharias discussing research with colleague Susan Larson in 1956.

Leona Zacharias died in 1990, just shy of her eighty-third birthday. I'll always treasure the interlude of intimacy I had with her during my childhood. And now I can add to it the memory of the intense connection I felt during those days I spent reading through her files. For the first time, I felt like I was seeing all of her, and she was inviting me into her brilliant mind.

# SELECTED BIBLIOGRAPHY

The lives and accomplishments of the female scientists in this book would not have come to light without the tremendous work, passion, and curiosity of fellow researchers and experts, including those at the Lost Women of Science Initiative. Listen to the award-winning *Lost Women of Science* podcast to continue learning about the innovative women in this book as well as others whose stories are still being told.

If you would like to keep exploring, here is a selection of websites, books, archives, and museums, all of which help ensure that the contributions of these women will be lost no more.

## Websites


lostwomenofscience.org

scientificamerican.com

womenshistory.org


### Eunice Newton Foote


climate.gov/news-features/features/happy-200th-birthday-eunice-foote-hidden-climate-science-pioneer

nytimes.com/2020/04/21/obituaries/eunice-foote-overlooked.html

scientificamerican.com/article/the-woman-who-demonstrated
-the-greenhouse-effect

**Flora Patterson**

apsnet.org/edcenter/apsnetfeatures/Pages/FloraPatterson.aspx
scientificamerican.com/article/this-overlooked-scientist-helped
-save-washington-d-c-s-cherry-trees
si.edu/object/siris_arc_306391

**Dr. Sarah Loguen Fraser**

scientificamerican.com/article/reconstruction-helped-her
-become-a-physician-jim-crow-drove-her-to-flee-the-u-s/
syracuse.com/news/2003/02/sarah_loguen_fraser.html

**Elizebeth Smith Friedman**

nsa.gov/History/Cryptologic-History/Historical-Figures/Historical
-Figures-View/Article/1623028/elizebeth-s-friedman
womenshistory.org/education-resources/biographies
/elizebeth-smith-friedman

**Dr. Cecilia Payne-Gaposchkin**

amnh.org/learn-teach/curriculum-collections
/cosmic-horizons-book/cecilia-payne-profile
amphilsoc.org/blog/cecilia-payne-gaposchkin-1900-1979

**Dr. Dorothy Andersen**

cfmedicine.nlm.nih.gov/physicians/biography_8.html

library-archives.cumc.columbia.edu/finding-aid
/dorothy-h-andersen-papers-1930-1966

**Klára Dán von Neumann**

discover.lanl.gov/news/0323-von-neumanns-letters

smithsonianmag.com/science-nature/meet-computer-scientist
-you-should-thank-your-phone-weather-app-180963716

**Dr. Isabella Aiona Abbott**

lowellmilkencenter.org/programs/projects/view/dr-isabella
-abbott-first-lady-of-limu/hero

pbshawaii.org/long-story-short-with-leslie-wilcox-isabella
-aiona-abbott/

**Yvonne Y. Clark**

alltogether.swe.org/2022/02/y-y-clarks-legacy

news.vanderbilt.edu/2019/05/23/yvonne-young-clark-ms72
-first-lady-of-engineering

scientificamerican.com/video/the-first-lady-of-engineering-lost
-women-of-science-podcast-season-3-episode-1

www.tnstate.edu/engineering/drclark.aspx

**Leona Zacharias**

focus.masseyeandear.org/uncovering-the-work-of-a-pediatric-eye-disease-pioneer

scientificamerican.com/article/leona-zacharias-helped-solve-a-blindness-epidemic-among-premature-babies-she-received-little-credit

## Books


*Breath from Salt: A Deadly Genetic Disease, a New Era in Science, and the Patients and Families Who Changed Medicine Forever* by Bijal D. Trivedi

*Dorothy Hansine Andersen: The Life and Times of the Pioneering Physician-Scientist Who Identified Cystic Fibrosis* by John Scott Baird

*ENIAC in Action: Making and Remaking the Modern Computer* by Thomas Haigh, Mark Preistley, and Crispin Rope

*The Fire of Stars: The Life and Brilliance of the Woman Who Discovered What Stars Are Made Of* by Kristen W. Larson, illustrated by Katherine Roy

*The Glass Universe: How the Ladies of the Harvard Observatory Took the Measure of the Stars* by Dava Sobel

*Turing's Cathedral: The Origins of the Digital Universe* by George Dyson

*What Stars Are Made Of: The Life of Cecilia Payne-Gaposchkin* by Donovan Moore

*The Woman Who Smashed Codes: A True Story of Love, Spies, and the Unlikely Heroine Who Outwitted America's Enemies* by Jason Fagone

*Yvonne Clark and Her Engineering Spark* by Allen R. Wells, illustrated by DeAndra Hodge

## Museums

American Museum of Natural History

Computer History Museum

Library of Congress

National Museum of African American History and Culture

National Women's History Museum

# IMAGE CREDITS

p. 4: Lionel Pincus and Princess Firyal Map Division, The New York Public Library. "Underground routes to Canada"/New York Public Library Digital Collections; p. 5: "Troy Female Seminary," 1800, Broadsides, leaflets, and pamphlets from America and Europe, Library of Congress, 2020771134; p. 6: "Our Roll of Honor," May 1908, Miller NAWSA Suffrage Scrapbooks, 1897–1911, Library of Congress, JK1881 .N357 sec. XVI, no. 3–9 NAWSA Collection; p. 13: *The American Journal of Science and Arts*: Second Series: Vol. XXII–November 1856, Missouri Botanical Garden Library, 1856; p. 14: The Bakerian Lecture, "Tyndall's Setup for Measuring Radiant Heat Absorption by Gases Annotated"/Wikimedia Commons; p. 23: "Harvard Botanic Garden looking toward the Gray House, Gray Herbarium, and Conservatories, circa 1890–1893." ID 1171. Botany Libraries photograph collection, circa 1770–2020. gra00083. Archives of the Gray Herbarium, Harvard University; p. 24: Flora W. Patterson, "A Collection of Economic and Other Fungi Prepared for Distribution," 1902, United States Bureau of Plant Industry/Wikimedia Commons; p. 26: "Omphalotus Olearius," June 12, 2011/ Wikimedia Commons; p. 30: Scott Bauer, "Washington C D.C. Tidal Basin Cherry Trees," April 1999, United States Department of Agriculture/Wikimedia Commons; p. 32: "Beagle Brigade," March 14, 2001, US Department of Agriculture/Wikimedia Commons; p. 34: "Mrs. F. W. Patterson," c. 1910–1920, National Photo Company Collection, Library of Congress, LC-F82-10259; p. 39: "The Plot Exploded!," June 4, 2014, Boston Public Library/Wikimedia Commons; p. 41: Benjamin Powelson, "Portrait of Harriet Tubman," c. 1868–1869, Emily Howland Photograph Album, Library of Congress, 2018645050; p. 46: Mathew B. Brady, "Fred K. Douglas [i.e. Frederick Douglass] / Brady, Washington, D.C.," c. 1880, Library of Congress; p. 47: "Fraser Family's Pharmacy in the Puerto Plata, Dominican Republic," Colored Conventions/Wikimedia Commons; p. 52: "Linda Brown Smith, Ethel Belton Brown, Harry Briggs Jr., and Spottswood Bolling Jr. During Press Conference at Hotel Americana," June 9, 1964, *New York World Telegram* and the *Sun* Newspaper, Library of Congress, 95503560; p. 54: "Tintype of a woman carrying a medical bag," c. 1890s, National Museum of African American History and Culture, 2014.37.14; p. 59: "Elizebeth Smith Friedman," National Security Agency/Central Security Service; p. 61 (top left): "Image 1 of The Keys for Deciphering the Greatest Works of Sir Francis Bacon, Baron of Verulam, Viscount St. Alban," Riverbank Laboratories, Library of Congress, 17004326; p. 61 (top right and bottom): "Image

39 of The Keys for Deciphering the Greatest Works of Sir Francis Bacon, Baron of Verulam, Viscount St. Alban," Riverbank Laboratories, Library of Congress, 17004326; p. 61: loc.gov/resource/gdcmassbookdig.keysfordecipheri00rive/?sp=39&r=-0.809,-0.152,2.618,1.417,0; p. 72: "Medal Presentation Smithsonian Code Talkers Bush," July 26, 2001, White House Photo Office/Wikimedia Commons; p. 73: "Muzeum 2 Wojny Swiatowej Gdansk Enigma Cipher Machine," April 11, 2017/ Wikimedia Commons; p. 79: "Cecilia Helena Payne," 1904, Harvard Square Library, courtesy of Katherine Haramundanis; p. 81: "Image of Cecilia Payne-Gaposckin," Smithsonian Archives, courtesy of Project PHaEDRA, SIA2009-1327; p. 84: A. Sonrel, "Great Refractor," Harvard College Observatory/Wikimedia Commons; p. 86: "Observatory Women [photographic group portrait, 1925," 1925, Harvard University Archives HUPSK Observatory (19), photo order number 11645; p. 90: "Katherine, Martin Schwardschild, Cecilia, Otis Scammon, Peter, Sergei, and Edward," 1941, Harvard Square Library courtesy of Katherine Haramundanis; p. 100: "Mount Holyoke College," Boston, 1908, Library of Congress, 2018756719; p. 104: "Babies Hospital, Medical Center, B'Way & 168th St., New York," 1930, Library of Congress, 2024691130; p. 109: "Dorothy H. Andersen, Professor of Pathology," c. 1960, P&S: The Yearbook of the College of Physicians and Surgeons, Columbia University/Wikimedia Commons; p. 112: "Matilda Joslyn Gage"/Wikimedia Commons; p. 120: Alfred T. Palmer, "The More Women at Work the Sooner We Win!" US Government Printing Office, 1943, Library of Congress, LC-USZ62-112283; p. 122: "John von Neumann Los Alamos Identity Badge Photo," c. 1944, Los Alamos National Laboratory/Wikimedia Commons; p. 124: "Reprogramming ENIAC," 1946, ARL Technical Library/Wikimedia Commons; p. 129: © Ralf Roletschek, "Spielbank-Wiesbaden (roulette wheel)," February 27, 2013/Wikimedia Commons; p. 138: "Agassiz Building, Hopkins Marine Station," July 5, 2015/Wikimedia Commons; p. 140: "Hypnea musciformis (herbarium item)," July 1982/Wikimedia Commons; p. 156: Russell Lee, "'Colored' Drinking Fountain from Mid-20th Century with African-American Drinking," July 1939, Library of Congress, 2017740552; p. 159: "The Campus, Howard University," 1910/Wikimedia Commons; p. 162: "Apollo 11 Launch," July 16, 1969, NASA/Wikimedia Commons; p. 169: Image of Dr. Leona Zacharias courtesy of Johanna Zacharias; p. 172: "Harvard Medical School, Boston, Mass," c. 1930–1945, Boston Public Library Tichnor Brothers Collection/ Wikimedia Commons; p. 173: "Beckman Model D Oxygen Meter in Use with an Infant's Incubator," c. 1959, Science History Institute/Wikimedia Commons; p. 179: Ralph_PH, "SWonderBSTH," July 6, 2019/Wikimedia Commons; p. 183: James F. Coyne. Courtesy of the Abraham Pollen Archives of Massachusetts Eye and Ear

# ACKNOWLEDGMENTS

This ambitious project would not have been possible without the help of a committed group of people. First there's the Lost Women of Science team, especially my co-founder, Amy Scharf, who hatched the plan to start Lost Women of Science with me. Then there are the individual producers who made the podcasts about the forgotten female scientists whose stories we tell in this book: Samia Bouzid, Luca Evans, Elah Feder, Ashraya Gupta, Barbara Howard, Dominique Janee, Zoe Kurland, Nora Mathison, Sophie McNulty, and Tracy Wahl.

Eowyn Burtner, our program manager, and Deborah Unger, our senior managing producer, both helped to get this project over the line, and Lexi Atiya and Jess Gingrich were our great fact-checkers.

And because at Lost Women of Science we want to make sure everything is accurate, we want to thank advisory board member Dr. Ellen Lyon for her help reviewing the classroom experiments. A special shout-out to my aunt Johanna Zacharias, who trained her keen eye on the chapter about my grandmother Leona Zacharias.

We owe a huge debt of gratitude to literary agent Jim Levine, who's game for pretty much anything, and this project was no exception. And big thanks to the talented and dedicated team at Bright Matter Books and Penguin Random House, especially our editor, Elizabeth Stranahan. Thanks, too, to Tom Russell, who loved the idea from the start, as well as managing editor Rebecca Vitkus, copy editors Alison Kolani and Maddy Stone, cover designer Carol Ly, and illustrator Karyn Lee.

Finally, our very special thanks go to Melina Gerosa Bellows, who took our podcasts and transformed them into stories on a page.

## ABOUT THE LOST WOMEN OF SCIENCE INITIATIVE

**THE LOST WOMEN OF SCIENCE INITIATIVE** is a nonprofit with two overarching and interrelated missions: to tell the story of female scientists who made groundbreaking achievements in their fields yet remain largely unknown to the general public, and to inspire girls and young women to pursue education and careers in STEM. To uncover more pioneering women in science, listen to their award-winning podcast or visit the Lost Women of Science online.

lostwomenofscience.org

# ABOUT THE AUTHOR

**KATIE HAFNER** is the host and co-creator of the award-winning *Lost Women of Science* podcast. She was a longtime reporter for the *New York Times*, which she still contributes to, and has written for the *New York Times Magazine*, *Esquire*, *Wired*, and *O*, among other outlets. She has covered women in STEM for more than thirty years. She lives in San Francisco with her husband.

katiehafner.com

# ABOUT THE AUTHOR

**MELINA GEROSA BELLOWS** is the president of Fun Factory Press, a boutique publishing business specializing in children's nonfiction content. She has created hundreds of new products and franchises, including the *New York Times* bestselling National Geographic Kids Almanac, Weird but True!, the reintroduction of *National Geographic Kids* magazine, and the Totally Random Facts series. She lives in Washington, DC, with her two children and numerous pets.

melinabellows.com

# ABOUT THE ILLUSTRATOR

**KARYN LEE** is an illustrator and designer residing in her native home of New York City. She can also be found designing children's books during the day, listening to pop culture podcasts, and using her limited scientific knowledge to bake all kinds of sweets.

karynslee.com